Learning Statistics Through Playing Cards

To Helen

Learning Statistics Through Playing Cards

Thomas R. Knapp

SAGE Publications
International Educational and Professional Publisher
Thousand Oaks   London   New Delhi

For information address:

SAGE Publications, Inc.
2455 Teller Road
Thousand Oaks, California 91320
E-mail: order@sagepub.com

SAGE Publications Ltd.
6 Bonhill Street
London EC2A 4PU
United Kingdom

SAGE Publications India Pvt. Ltd.
M-32 Market
Greater Kailash I
New Delhi 110 048 India

Printed in the United States of America

Library of Congress Cataloging-in-Publication Data

Knapp, Thomas R., 1930-
    Learning statistics through playing cards / author, Thomas R.
    Knapp.
        p.   cm.
    Includes bibliographical references and index.
    ISBN 0-7619-0108-6 (acid-free paper). — ISBN 0-7619-0109-4 (pbk: acid-free paper)
    1. Statistics.   I. Title.
QA276.12.K597 1996
519.5—dc20   95-41790

This book is printed on acid-free paper.

96 97 98 99 00 10 9 8 7 6 5 4 3 2 1

Sage Production Editor: Diane S. Foster

Contents

Tables and Figures

Preface

I believe that all of the important concepts in statistics can be taught by using an ordinary deck of playing cards. This little book is an outgrowth of that belief. It is intended to serve as a principal or supplementary textbook for three different audiences: (a) those who have already taken a statistics course and either have failed or have gotten so little out of it that they want to start all over again, (b) those who have already taken a statistics course and have done quite well but want to sharpen their understanding of the basic concepts, and (c) those who are studying statistics for the first time and are attracted by the idea of using playing cards for something other than games.

The nine chapters cover many of the topics that are included in a one-quarter or one-semester college course at the introductory, "non-calculus" level. There are no algebraic formulas whatsoever (the general approach is very verbal—I want my readers to be able to *speak* the language of statistics). Modest competence in the four basic arithmetic operations (addition, subtraction, multiplication, and division) and square-root extraction is sufficient mathematical preparation.

A word about calculators and computers: Although I would prefer that you work through most or all of the exercises at the end of the chapters "by hand," if you find that to be either boring or unduly difficult I have no objection to your using some sort of computational assistance (machines, not friends).

Why should people learn statistics? First of all, the popular press and other media are full of statistics (particularly percentages and differences between percentages, which are the primary focus of this book), and an understanding of basic statistical concepts helps immeasurably in trying to sort out what reported statistics to regard seriously and what to take with a grain of salt. Second, it is almost impossible to read the scientific research literature, much less carry out such research, without a knowledge of statistics. Finally, like mathematics (its parent discipline), statistics is based upon sound logical principles. When properly used, statistical analysis actually makes sense.

Why use a deck of cards? The principal reason is that a deck of cards constitutes an actual finite population from which samples can be randomly drawn, both with replacement and without replacement. If this book emphasized the use of coins, for example, the population would be obscure, hypothetical, and infinite (all possible tosses of a fair coin), and sampling would necessarily be with replacement only.

So get out your deck of cards, turn to page 1, and begin. I think you might even like it.

Thomas R. Knapp
Columbus, Ohio

Populations, Variables, and Distributions

Do you have your deck of cards? If so, spread them out and take a look at them. You probably already know that

1. there are 52 of them; and
2. the deck is made up of two different colors (black and red), four different suits (clubs, diamonds, hearts, and spades; the first and last of which are black, and the middle two of which are red), 13 different denominations (ace [1], 2, 3, 4, 5, 6, 7, 8, 9, 10, jack [11], queen [12], and king [13]), and two different types of "pictureness" (face cards [jacks, queens, and kings] and non-face cards [all others]).

Table 1.1 contains a list of the names of all of the 52 cards. In statistics, any entire collection of objects is called a *population*.

The characteristics of the playing cards (color, suit, denomination, and pictureness) are called *variables*, because the cards are not all of the same color, suit, denomination, or pictureness. The values

Table 1.1   The Objects in the Population of Cards

1. The ace of clubs	27. The ace of hearts
2. The 2 of clubs	28. The 2 of hearts
3. The 3 of clubs	29. The 3 of hearts
4. The 4 of clubs	30. The 4 of hearts
5. The 5 of clubs	31. The 5 of hearts
6. The 6 of clubs	32. The 6 of hearts
7. The 7 of clubs	33. The 7 of hearts
8. The 8 of clubs	34. The 8 of hearts
9. The 9 of clubs	35. The 9 of hearts
10. The 10 of clubs	36. The 10 of hearts
11. The jack of clubs	37. The jack of hearts
12. The queen of clubs	38. The queen of hearts
13. The king of clubs	39. The king of hearts
14. The ace of diamonds	40. The ace of spades
15. The 2 of diamonds	41. The 2 of spades
16. The 3 of diamonds	42. The 3 of spades
17. The 4 of diamonds	43. The 4 of spades
18. The 5 of diamonds	44. The 5 of spades
19. The 6 of diamonds	45. The 6 of spades
20. The 7 of diamonds	46. The 7 of spades
21. The 8 of diamonds	47. The 8 of spades
22. The 9 of diamonds	48. The 9 of spades
23. The 10 of diamonds	49. The 10 of spades
24. The jack of diamonds	50. The jack of spades
25. The queen of diamonds	51. The queen of spades
26. The king of diamonds	52. The king of spades

for the variables are called *observations* or *measurements*. The measurements on each of the four variables for each of the 52 cards are contained in Table 1.2. This rectangular array of numbers, where each row (horizontal) represents an object and each column (vertical) represents a variable, is called a *data matrix*. In statistics we usually have many more objects than we have variables, so most data matrices are "long and skinny" rather than "short and fat."

As indicated above, there are just two categories for the color variable (black and red) and for the pictureness variable (face card and non-face card). Any variable that has just two categories is called a *dichotomy*. For such variables it is often convenient to use numbers to identify the categories even though the numbers may have no necessary relevance to the categories. The numbers most often chosen for this purpose are 0 and 1, in which case the dichotomy

Table 1.2 The Data Matrix for the Population of Cards

Object	Color	Suit	Denomination	Pictureness
1	1	1	1	0
2	1	1	2	0
3	1	1	3	0
4	1	1	4	0
5	1	1	5	0
6	1	1	6	0
7	1	1	7	0
8	1	1	8	0
9	1	1	9	0
10	1	1	10	0
11	1	1	11	1
12	1	1	12	1
13	1	1	13	1
14	0	2	1	0
15	0	2	2	0
16	0	2	3	0
17	0	2	4	0
18	0	2	5	0
19	0	2	6	0
20	0	2	7	0
21	0	2	8	0
22	0	2	9	0
23	0	2	10	0
24	0	2	11	1
25	0	2	12	1
26	0	2	13	1
27	0	3	1	0
28	0	3	2	0
29	0	3	3	0
30	0	3	4	0
31	0	3	5	0
32	0	3	6	0
33	0	3	7	0
34	0	3	8	0
35	0	3	9	0
36	0	3	10	0
37	0	3	11	1
38	0	3	12	1
39	0	3	13	1
40	1	4	1	0
41	1	4	2	0
42	1	4	3	0
43	1	4	4	0
44	1	4	5	0

(continued)

Table 1.2   Continued

Object	Color	Suit	Denomination	Pictureness
45	1	4	6	0
46	1	4	7	0
47	1	4	8	0
48	1	4	9	0
49	1	4	10	0
50	1	4	11	1
51	1	4	12	1
52	1	4	13	1

is called a *dummy variable.* For the color variable we will arbitrarily call all of the black cards 1s and all of the red ("non-black") cards 0s; for the pictureness variable we will call all of the face cards 1s and all of the non-face cards 0s. In this book we will pay special attention to dichotomies and to the statistical procedures that are appropriate for dealing with them.

For the suit variable we will use the numbers 1, 2, 3, and 4 to identify the clubs, diamonds, hearts, and spades, respectively, because these are the rank orders (from lowest to highest) of the four suits in the game of bridge.

The best way to get a feel for the concept of variables is to scrutinize carefully two cards that are about as different as they can possibly be. Let us therefore make a "case study" of the queen of diamonds (card 25 in the population) and the 6 of spades (card 45). The color of the queen of diamonds is red (0), its suit is diamonds (2), its denomination is 12 (12), and it is a face card (1). The six of spades is black (1) rather than red, it is a spade (4) and not a diamond, it has a denomination of 6 (6), and it is not a face card (0).

Frequency Distributions

The first thing you should do whenever you have the data for a particular variable for a population is to make a *frequency distribution* of those data. A frequency distribution is nothing more than a count of the number of times each value of the variable occurs. I'm sure you've made a lot of frequency distributions in your lifetime, but you probably have never called them by that name. What did you do? You wrote down all of the possible values in a column, put tally marks in the appropriate places as you checked each value off a list, and then counted the tallies. If you were to do that for each of the four playing card variables, you would get the frequency distributions displayed in Table 1.3.

The tally sections of the frequency distributions in Table 1.3 provide graphic representations of those distributions. If you rotate your book 90 degrees counter-clockwise, the tallies form what is called a *histogram,* with the values of the variable along the horizontal (x) axis and the frequencies along the vertical (y) axis.

In addition to the "raw" frequencies, Table 1.3 also contains the *relative frequencies* and corresponding percentages. For the pictureness variable, for example, the frequency for face card is 12 and the relative frequency is 12 out of 52, or .231, which converts to 23.1% (decimal fractions may be easily converted to percentages by moving the decimal point two places to the right and affixing a % sign).

Three of the distributions (color, suit, and denomination) are *symmetric*—that is, they are perfectly balanced—but the fourth distribution (pictureness) is *asymmetric* or *skewed.* It is often interesting to summarize certain features of frequency distributions; these features—central tendency, variability, skewness, and kurtosis— will be pursued in Chapter 2.

Children and States

Although playing cards constitute an ideal population for learning basic statistical concepts, we need more interesting populations to which we might apply those concepts. I have therefore created two other populations of the same size but composed of different "objects." The first of these is a hypothetical group of 52 children who attend a hypothetical private nursery school. The names of the children (names I took from the Rochester, New York, telephone directory) are listed in Table 1.4. Donna Abbey is equivalent to the ace of clubs, Julia Ayres is like the 2 of clubs, and so on, down to

Table 1.3 The Frequency Distributions of the Four Variables for the Population of Cards

Variable	Value	Tally	Frequency	Relative Frequency
Color				
	0	1111111111111111111111111	26	.500 (50%)
	1	1111111111111111111111111	26	.500 (50%)
	total		52	
Suit				
	1	1111111111111	13	.250 (25%)
	2	1111111111111	13	.250 (25%)
	3	1111111111111	13	.250 (25%)
	4	1111111111111	13	.250 (25%)
	total		52	
Denomination				
	1	1111	4	.077 (7.7%)
	2	1111	4	.077 (7.7%)
	3	1111	4	.077 (7.7%)
	4	1111	4	.077 (7.7%)
	5	1111	4	.077 (7.7%)
	6	1111	4	.077 (7.7%)
	7	1111	4	.077 (7.7%)
	8	1111	4	.077 (7.7%)
	9	1111	4	.077 (7.7%)
	10	1111	4	.077 (7.7%)
	11	1111	4	.077 (7.7%)
	12	1111	4	.077 (7.7%)
	13	1111	4	.077 (7.7%)
	total		52	

Pictureness

0   1111111111111111111111111111111111111111   40   .769 (76.9%)
1   111111111111   12   .231 (23.1%)

total   52

Table 1.4   The Objects in the Population of Children

1. Donna Abbey	27. Bruce Naab
2. Julia Ayres	28. Wasyl Nyznk
3. Edna Baars	29. Kack Oagley
4. Louise Byron	30. Robert Ozols
5. Ana Cabera	31. Daniel Paap
6. Marylou Czudak	32. John Pytlak
7. Mabel Daansen	33. Duane Quackenbush
8. Florence Dzierzanowski	34. Richard Quodomine
9. Ethel Eaglestone	35. Charles Raab
10. Ruth Ezell	36. Richard Rzepkowski
11. Marion Faas	37. Helmut Saager
12. Irene Futherer	38. Bruce Szypot
13. Nancy Gabbey	39. Luigi Tabacco
14. David Gzik	40. Peggy Tylter
15. Chester Ha	41. Anna Uberty
16. Walter Hyz	42. Lillian Utz
17. John Iabone	43. Camilla Vacca
18. Vincent Izzo	44. Anna Vroman
19. Francis Jabout	45. Laura Wachsman
20. Joseph Juzwiak	46. Kate Wysk
21. Alex Kaalinagy	47. Alexandra Xanthos
22. Richard Kyte	48. Jean Xydias
23. William Laack	49. Carmela Yacco
24. Scott Lyzwa	50. Catherine Yurenovich
25. Carl Maar	51. Mary Zabkar
26. Richard Mytych	52. Alice Zutterman

Alice Zutterman, who is like the king of spades. Even though these are "faked" data, I think you will find this population to be reasonably realistic, except for the fact that all children with last names beginning with the letters A to F are girls, there are many Polish names, and so on (perhaps the school has a strange admissions policy).

The second of the populations I have created is a collection of the states of the United States (plus the District of Columbia and Puerto Rico, to make a total of 52). The names of the states are listed in Table 1.5 (although 2 of these 52 are not states, strictly speaking, I will refer to this population throughout as the 52 states). Delaware is equivalent here to the ace of clubs, Pennsylvania is like the 2 of clubs, and so on, down to Puerto Rico, which is like the king of spades. The order in which the states are listed is a bit unusual.

Table 1.5 The Objects in the Population of States

1. Delaware	27. California
2. Pennsylvania	28. Oregon
3. New Jersey	29. Nevada
4. Georgia	30. Colorado
5. Connecticut	31. Montana
6. Massachusetts	32. Washington
7. Maryland	33. Idaho
8. South Carolina	34. Wyoming
9. New Hampshire	35. Utah
10. Virginia	36. New Mexico
11. New York	37. Arizona
12. North Carolina	38. Alaska
13. Rhode Island	39. Hawaii
14. Louisiana	40. Vermont
15. Illinois	41. Kentucky
16. Missouri	42. Tennessee
17. Arkansas	43. Ohio
18. Texas	44. Indiana
19. Iowa	45. Mississippi
20. Wisconsin	46. Alabama
21. Minnesota	47. Maine
22. Kansas	48. Michigan
23. Nebraska	49. Florida
24. North Dakota	50. West Virginia
25. South Dakota	51. District of Columbia
26. Oklahoma	52. Puerto Rico

The first 13 are the original 13 colonies, but please don't ask me why the rest of the states are listed in the particular order shown.

At the end of each of the first eight chapters in this book, there is a series of exercises based on these alternative populations. Here is the first set. (The answers to most of the exercises are provided at the back of the book, but be sure you work on the exercises *before* you peek at the answers.)

Exercises

1. Three of the nursery school children have first initials that are the same as their last initials. Which children are these? Would that characteristic (same or different initials) be classified as a dichotomy? Why or why not? Would its frequency distribution for this population be symmetric or skewed? Why?

2. Suppose there were a variable called "score on speaking vocabulary test" that had the same frequency distribution for the nursery school children as the denomination variable for the deck of cards. Which four children obtained the highest score on that test? Which four had the lowest score?

3. Which of the 52 states has the most letters in its name? Which has the fewest? Make a frequency distribution for the variable "number of letters in name." Is it symmetric or skewed? Why?

Table 1.6   The Number of Members in the U.S. House of Representatives From Each State

1.	Delaware	1	27.	California	52
2.	Pennsylvania	21	28.	Oregon	5
3.	New Jersey	13	29.	Nevada	2
4.	Georgia	11	30.	Colorado	6
5.	Connecticut	6	31.	Montana	1
6.	Massachusetts	10	32.	Washington	9
7.	Maryland	8	33.	Idaho	2
8.	South Carolina	6	34.	Wyoming	1
9.	New Hampshire	2	35.	Utah	3
10.	Virginia	11	36.	New Mexico	3
11.	New York	31	37.	Arizona	6
12.	North Carolina	12	38.	Alaska	1
13.	Rhode Island	2	39.	Hawaii	2
14.	Louisiana	7	40.	Vermont	1
15.	Illinois	20	41.	Kentucky	6
16.	Missouri	9	42.	Tennessee	9
17.	Arkansas	4	43.	Ohio	19
18.	Texas	30	44.	Indiana	10
19.	Iowa	5	45.	Mississippi	5
20.	Wisconsin	9	46.	Alabama	7
21.	Minnesota	8	47.	Maine	2
22.	Kansas	4	48.	Michigan	16
23.	Nebraska	3	49.	Florida	23
24.	North Dakota	1	50.	West Virginia	3
25.	South Dakota	1	51.	District of Columbia	0
26.	Oklahoma	6	52.	Puerto Rico	0

4a. The 1995 edition of the *Information Please Almanac* lists the names of the members of the U.S. House of Representatives for each state (the District of Columbia and Puerto Rico have none, but zero is a perfectly respectable measurement). The numbers of members from each state are provided in Table 1.6 and recorded on the map in Figure 1.1. Make a frequency distribution for that variable. Comment on any interesting features this distribution may have.

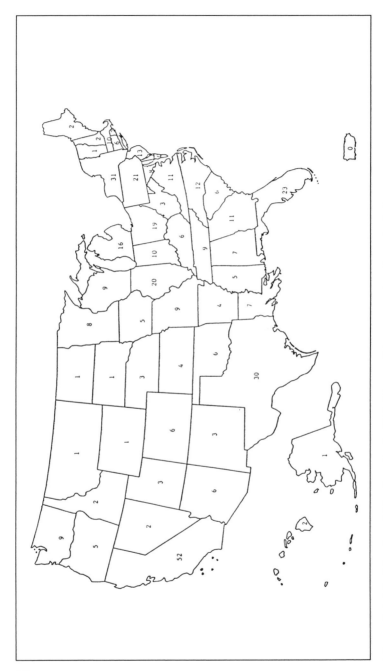

Figure 1.1. The Number of Members in the U.S. House of Representatives From Each State

Table 1.7 The Political Affiliations of Members in the U.S. House of Representatives From the State of California

District	Affiliation	District	Affiliation
1	Republican	27	Republican
2	Republican	28	Republican
3	Democratic	29	Democratic
4	Republican	30	Democratic
5	Democratic	31	Democratic
6	Democratic	32	Democratic
7	Democratic	33	Democratic
8	Democratic	34	Democratic
9	Democratic	35	Democratic
10	Republican	36	Democratic
11	Republican	37	Democratic
12	Democratic	38	Republican
13	Democratic	39	Republican
14	Democratic	40	Republican
15	Democratic	41	Republican
16	Democratic	42	Democratic
17	Democratic	43	Republican
18	Democratic	44	Republican
19	Republican	45	Republican
20	Democratic	46	Republican
21	Republican	47	Republican
22	Republican	48	Republican
23	Republican	49	Republican
24	Democratic	50	Democratic
25	Republican	51	Republican
26	Democratic	52	Republican

4b. The *Information Please Almanac* also provides information on the political affiliation of each of the members of the House of Representatives from the state of California—see Table 1.7 (by a happy coincidence, there just happen to be 52 of them also). How many of those members are Republicans?

5. As you may know, the primary purpose of the decennial census is to allow for fair apportionment of representation for the states in the U.S. House of Representatives. The frequency distribution for the variable "number of members in the House of Representatives" is therefore dynamic rather than static, even though the total frequency stays the same (435). What do you think the distribution will look like in the year 2020? Why?

Parameters

It is often cumbersome, and unnecessary, to preserve all of the information contained in a frequency distribution for a population. We will therefore concentrate on a few indexes called *parameters* that summarize the important features of a population distribution. But what features should we emphasize? Karl Pearson and other statisticians have suggested that there are four things we usually want to know about a given distribution:

1. Its *central tendency* (i.e., some sort of average measurement for the variable)
2. Its *variability* or *dispersion* (i.e., some indication of the extent to which the measurements differ from one another)
3. Its *skewness* (i.e., whether the distribution of the measurements is symmetric or skewed, and, if the latter, the direction and degree of asymmetry)
4. Its *kurtosis* (i.e., the degree to which the measurements tend to "pile up" at some point in the distribution)

Central Tendency

There are several popular indexes of central tendency, but the one that is used more often than all of the others put together is the *arithmetic mean*, or, more simply, the *mean*. To find the mean of a set of measurements, you add them up and divide by the number of them. You've been doing that all your life, haven't you?

The mean color (sounds funny, doesn't it?) of the 52 cards is the sum of the 26 1s and the 26 0s, which is 26, divided by 52, which is .50. This .50 is also the *proportion* of cards that are black (the 1s) and can of course be converted into a *percentage* (50%) by moving the decimal point two places to the right. Proportions and percentages are therefore special kinds of means. (*Note:* If symbols other than 0 and 1 are used to "code" the two colors, the mean color would *not* be the proportion of black cards.)

The other dichotomy, pictureness, has a mean of 12/52, or .231, or 23.1%—that is, 23.1% of the cards are picture cards. The means for each of the playing card variables are listed in Table 2.1 for each of the frequency distributions in Table 1.3.

Two other popular measures of central tendency are the *median* (the observation that divides the distribution in half) and the *mode* (the observation that occurs most frequently).

Variability

There are also several measures of variability. The easiest one, the *range*, is merely the difference between the lowest value and the highest value. The one that is used most often in scientific work, however, is the *standard deviation*. This is an index of dispersion around the arithmetic mean and is obtained as follows:

1. Find the mean. You already know how to do that.
2. Subtract the mean from each of the measurements. That's easy.
3. Square each of those differences; that is, multiply each difference by itself. That's easy, too, but remember that a plus times a plus is a plus, and a minus times a minus is also a plus.

Table 2.1 Descriptive Parameters for the Population of Cards

Variable	Mean	Standard Deviation	Skewness	Kurtosis
1. Color	.50	.50	0	1
2. Suit	2.5	1.118	0	1.640
3. Denomination	7.0	3.742	0	1.786
4. Pictureness	.231	.421	1.279	2.628

4. Add up all of those squared differences—also easy, albeit tedious.
5. Divide that sum by the number of measurements. This is the *variance*, which is the mean of the squared differences from the mean—if you follow that. (Authors of some statistics books say to divide the sum of the squared differences by *one less* than the number of measurements. The reasons for that are too complicated to go into here, so forget it.)
6. Take the square root of that quotient.

You were fine up until that last step, weren't you? What is the square root of a number? It is another number that when multiplied by itself gives you the number you started with. The square root of 4 is 2; that is, 2 times 2 is 4. The square root of 49 is 7, because 7 times 7 is 49, and so forth.

Let's work out the standard deviation of the color variable for the population of 52 cards:

1. The mean is .50, as previously calculated.
2. Zero minus .50 is −.50 for each of the 26 0s; 1 minus .50 is +.50 for each of the 26 1s.
3. The square of −.50 is +.25; so is the square of +.50.
4. The sum of those squares is 52 times .25, or 13.
5. Thirteen divided by 52 is .25 (the variance).
6. The square root of .25 is .50 (because .50 times .50 is .25). Therefore the standard deviation of the color variable is .50. (See Table 2.1 for this and for the standard deviations of the other variables.)

Got it? If not, don't be discouraged. You'll have lots of opportunities to practice the calculation of means and standard deviations when you work out the exercises at the end of this chapter.

Given that the standard deviation is the square root of the variance, and the variance is therefore the square of the standard deviation, choosing one of these two parameters over the other is largely a matter of personal preference. For theoretical work the variance has the simpler mathematical properties, but for applied work the standard deviation is used more frequently. The reason for this is that the standard deviation is in the "right" units, but the variance is in the "wrong" units. For example, if we had a population distribution of the number of eggs sold in various years by a dairy, the unit of measurement is eggs, the standard deviation comes out in eggs, but the variance comes out in squared eggs. (Do you see why? *Hint:* Study the six steps above very carefully and follow the unit of measurement to see where it gets squared and where it gets "unsquared.")

How do you interpret a standard deviation? It is best to think of the standard deviation as the "typical" difference between each measurement and the mean. This works particularly well for the color data; every measurement is a half unit away from the mean (the 0s are a half unit below the mean and the 1s are a half unit above the mean).

Skewness and Kurtosis

The other two features of a population distribution, skewness and kurtosis, are usually of considerably less interest than central tendency and variability, but indexes of those two properties can also be obtained, as follows.

SKEWNESS

1. Find the mean.
2. Subtract the mean from each measurement.

3. Find the cubes of those differences; that is, get the *third power* of the differences by multiplying each difference by itself, then by itself again. (*Note:* The cubes of the plus differences will be plus but the cubes of the minus differences will be minus. Do you see why?)
4. Add all of those up.
5. Divide by the total number of measurements; that is, find the mean of the cubed differences.
6. Divide *that* by the cube of the standard deviation.

KURTOSIS

Take the same six steps as above, but instead of cubing the differences and cubing the standard deviation, you raise them to the *fourth power* (i.e., you divide the mean of the fourth powers of the differences by the fourth power of the standard deviation).

I won't go through all of the calculations (see Table 2.1 for the answers), but the skewnesses of the distributions for the color, suit, and denomination variables are all equal to 0 (the cubes of the plus differences are "washed out" by the cubes of the minus differences). All symmetric distributions have a skewness of 0. Distributions whose histograms have a "hump" at the low end of the scale and a "tail" at the high end of the scale are called "skewed to the right" or "positively skewed"; their skewness is greater than 0. Distributions whose histograms have a "hump" at the high end and a "tail" at the low end are called "negatively skewed" or "skewed to the left." Although it has only two categories and the "hump" and the "tail" are not obvious, the pictureness variable is positively skewed.

The larger the kurtosis (anything over 3—which is the kurtosis of the bell-shaped or *normal* distribution—can be considered large), the more the measurements tend to pile up around a single point. Distributions with a kurtosis greater than 3 are sometimes called *leptokurtic*; those with a kurtosis less than 3 are called *platykurtic*. Because half of the colors are 0s and the other half are 1s, that distribution has a very low kurtosis (as you can see in Table 2.1, it is actually equal to 1).

Some authors suggest that you subtract 3 from the final result of the kurtosis calculation so that the normal distribution will have a kurtosis of 0 and the kurtosis of all other distributions can be evaluated with respect to 0 rather than 3. Many computer programs have incorporated this recommendation.

A final note about the word *parameter*, which has at least two different meanings in the nonstatistical world. The first is synonymous with *dimension*, as in, What are the parameters of this problem? The second is synonymous with *boundary*, as in, Within what parameters are we permitted to operate? (This appears to me to be a confusion with the word *perimeter*.) Please try to suppress both of those meanings, at least as far as this book is concerned.

Exercises

1. Find the mean and the standard deviation of the vocabulary test scores for the 52 nursery school children (see Exercise 2 at the end of Chapter 1).

2a. What is the range of the vocabulary test score distribution?

2b. It can be shown that the standard deviation can be no larger than one-half of the range and no smaller than the range divided by the square root of twice the number of observations, regardless of the "shape" of the distribution. Use this fact to check your calculation of the standard deviation of the vocabulary test scores.

3. Calculate the mean and the standard deviation for the number of members in the House of Representatives variable for the 52 states. (Use the actual data in Table 1.6, but see Exercise 4a at the end of Chapter 1.)

4. Calculate the skewness and the kurtosis for the same distribution as in Exercise 3. Do those numbers make sense? Why or why not?

5. You can compare the means and standard deviations of two distributions if the distributions have the same scale, in order to determine which distribution is "shoved over farther to the right" and /or which is more "spread out," but it is not appropriate to compare the means and standard deviations of two distributions if they have different scales. Why is that?

Percentages

The only parameters we shall be concerned with for the rest of this book are percentages and differences between percentages, for reasons of (a) simplicity and (b) ubiquity: Percentages are generally easier to calculate and understand than other indexes, and they come up all the time.

Cautions

Just about everybody knows what a percentage is and how to calculate one. A percentage is a measure of a part of a whole, and is calculated by dividing the number of things in the part by the number of things in the whole and then multiplying by 100. For example, in the discussion in Chapter 1 of the pictureness variable for the population of playing cards, I pointed out that the number of face cards in a deck of cards is 12 out of 52, or .231, or 23.1%. Percentages can be tricky, however, so a few cautions are in order.

First, percentages corresponding to each of the parts in a whole must add to 100. That may seem obvious, but it is surprising how often they don't in actual research reports. One reason is rounding error, which can be remedied by carrying out the calculations to a larger number of decimal places—although this can be annoying (for instance, is 20 out of 30 66%, 67%, 66.6%, 66.7%, 66.66%, or 66.67%?). Another reason has to do with situations in which there is overlap, such as overlapping groups of patients suffering from various ailments. The percentages of those with lung cancer plus those with AIDS plus those with hypertension might very well add to *more than 100* if some patients are diagnosed as having two or more of those problems. A third reason for failure to sum to 100 is missing data. If religious preference, say, is being analyzed, some subjects may not make such information available, and the percentages for the various religions will add to some number *less than 100*. They could be made to add to 100 if the number of non-missing data values, rather than the total sample size, were taken as the base, but this can be very confusing to the reader. It's best to include "missing" as an extra category.

The above reference to the base upon which percentages are calculated brings me to the second caution to be observed: Be careful of the changing base. There is an old joke about an employee who had to take a 50% decrease in salary from $400 a week to $200 a week, which the boss "restored" a month later by giving him a 50% increase. Because of the change in the base, the employee wound up with only $300 a week, not the original $400. In research, a common problem is that the investigator might try to compare a percentage for a *total* group at Time 1 with a percentage for a *surviving* group at Time 2. Suppose that in a longitudinal study of a particular birth cohort of elderly people (say, a group of people born in 1900), 5% had Alzheimer's disease at age 80 in 1980 but only 1% had Alzheimer's disease at age 90 in 1990. That doesn't mean that the individuals in the cohort got better. The base at age 80 may have been 1,000 whereas the base at age 90 may have been 700, with 43 of the original 50 Alzheimer's patients having died between age 80 and age 90.

A third caution has to do with the making of more than one comparison with percentages that have to add to 100. For example, if there is a difference of 30% between the percentage of Christians in good health and the percentage of non-Christians in good health, there must be a compensating difference in the opposite direction between the percentage of Christians in bad health and the percentage of non-Christians in bad health. A similar caution has to do with claims such as "80% of Christians are in good health, whereas only 20% are in bad health." If 80% are in good health, of course 20% are in non-good—that is, bad—health.

The final caution about percentages concerns very small bases. Percentages are both unnecessary and misleading when they are based on small sample sizes. (It should go without saying, but I'll say it anyhow, that the bases for percentages should *always* be specified.) If 80% of Christians but only 50% of non-Christians are reported to be in good health, that is no big deal if there are just 10 Christians and 10 non-Christians in the total sample, because it amounts to a difference of only 3 people.

Exercises

1. In Exercise 1 at the end of Chapter 1, you were asked how many of the nursery school children have first initials that are the same as their last initials. What *percentage* of the nursery school children have first initials the same as their last initials?

2. What percentage of the nursery school children scored more than one standard deviation above the mean on the speaking vocabulary test? (See Exercise 2 at the end of Chapter 1 and Exercise 1 at the end of Chapter 2.)

3. What percentage of the states have seven letters in their names?

4. What percentage of the states have fewer than 2 or more than 10 members in the House of Representatives? Are there any rounding problems in determining this percentage? Why or why not?

5. In Exercise 4b at the end of Chapter 1 you were asked to count the number of members in the House of Representatives from California who are Republicans. What *percentage* of those people are Republicans?

Probability and Sampling

There is much more to statistics than making frequency distributions for populations and summarizing various features of such distributions. As a matter of fact, we usually don't even have all of the observations for an entire population (for obvious practical reasons such as cost and time), so we can't actually construct the population distribution and calculate its parameters. What do we do? We take a *sample* of the population (i.e., a "piece" of the population), get some data for the sample, and try to say something about the population data that we wish we had. Does this sound confusing? Perhaps an example might help.

Suppose you had never seen a deck of cards before. Someone shows you one, intact and face down, and says to you: "Shuffle the deck, draw four cards from the deck, look at them, and make a guess as to what the percentage of black cards is for the whole deck." Take your deck of cards and do just that. I did, and the four cards I drew were cards 14, 29, 30, and 40 in our population—that is, the ace of diamonds, the 3 of hearts, the 4 of hearts, and the ace of spades. (What cards did you draw?) Because one out of the four cards in my

Table 4.1   A Frequency Distribution of Color for a Sample of Four Cards

Value	Tally	Frequency	Relative Frequency
0 (Non-black)	111	3	.750 (75%)
1 (Black)	1	1	.250 (25%)

sample is black (see Table 4.1 for the frequency distribution of the color variable for this sample), I would probably guess that 25% of the cards in the population are black. This would be wrong, of course, but that's the whole point—my sample consists of only a small portion of the population, so it would be unreasonable to expect that I would necessarily hit the correct percentage right on the button. (What would your guess be, based on your sample? Are you right or wrong?)

The process of generalizing from sample data, which we do know, to population data, which we don't know, is called *statistical inference*. The techniques for making such inferences are treated in Chapters 6 and 7; these are called *inferential statistics* (as opposed to *descriptive statistics*, which are the techniques for summarizing whatever data we happen to have in hand). An understanding of statistical inference depends upon a knowledge of both probability and sampling, however, to which I would now like to turn.

What Is Probability?

There are all kinds of fancy definitions of *probability* in the statistical literature, but one that is sufficient for our purposes is the following: *The probability of a particular outcome is its relative frequency among a specified set of outcomes.* You are already familiar with the concept of relative frequency, which was discussed in Chapter 1. In the playing card population the frequency of black (1), for example, is 26. The relative frequency of black is 26 divided by 52, or .50, given that there are 26 black cards out of a total of 52 cards. Therefore, if you were to shuffle the entire deck and draw one card, the probability is .50, or one chance in two, that it will be a black card. Let's try some other examples:

1. What is the probability of drawing a 9?
   Answer: 4/52 or .077
2. What is the probability of drawing a face card?
   Answer: 12/52 or .231
3. What is the probability of *not* drawing a face card?
   Answer: 40/52 or .769

Probabilities are numbers between 0 (impossibility) and 1 (certainty), and for any specified set of outcomes the respective probabilities must always add up to 1. For example, the probability of drawing a face card is 12/52 or .231; the probability of not drawing a face card is 40/52 or .769. Thus, 12/52 + 40/52 = 52/52, or 1 (in decimal form, .231 + .769 = 1). Therefore, if you know the probability that something will happen and you want to determine the probability that it won't happen, you can subtract the known probability from 1. Using this same example, the probability of not drawing a face card = 1 − .231 = .769.

Rules for Calculating Probabilities

There are two useful rules for calculating complex probabilities from simpler probabilities:

- *Rule 1* (the "and" rule): The probability that *both* of two outcomes will take place is the *product* of the probability that the first one will take place and the probability that the second one will take place, given that the first one took place (the so-called *conditional probability*).
- *Rule 2* (the "or" rule): The probability that *either* of two outcomes will take place is the *sum* of the probability that the first one will take place and the probability that the second one will take place, if the two outcomes cannot take place simultaneously.

These are a couple of mouthfuls, so let's take a couple of examples:

1. What is the probability of drawing two spades in two draws from a deck of cards, if the first card is replaced before the second card is drawn?
   Answer, by Rule 1: $13/52 \times 13/52 = 1/4 \times 1/4 = 1/16$, or .0667.
2. What is the probability of drawing two spades in two draws from a deck of cards, if the first card is *not* replaced before the second card is drawn?
   Answer, again by Rule 1: $13/52 \times 12/51 = 1/4 \times 4/17 = 1/17$, or .0588.

Sampling With or Without Replacement

In the first example above the probability that the second card is a spade does not depend on whether the first card was a spade (because the first card is replaced before the second one is drawn), so for each draw there are 52 cards that could be sampled and 13 of them are spades. In this case of *sampling with replacement*, the two outcomes ("spade on first draw" and "spade on second draw") are said to be *independent*. In the second example the probability that the second card is a spade does depend on whether the first card is a spade, because for the first draw there are 52 cards that could be sampled and 13 of them are spades, whereas for the second draw there are only 51 cards that could be sampled and only 12 of them are spades, *given that the first card is a spade*. In this case of *sampling without replacement*, the two outcomes are not independent. (Sampling with replacement essentially transforms a finite population into an infinite population, because there is always something left to sample.)

Now for some more examples:

3. What is the probability of drawing an ace or a king in a single draw?
   Answer, by Rule 2: $4/52 + 4/52 = 1/13 + 1/13 = 2/13$, or .154. (A single draw cannot yield a card that is both an ace and a king; those two outcomes are said to be *mutually exclusive*.)

4. What is the probability of drawing two black cards and two red cards in four draws from a deck of cards, *without replacement*? Answer, by extensions of Rule 1 *and* Rule 2 (hold on to your hats for this one): Possible "favorable" outcomes for this problem are all permutations of the form BBRR (B = black, R = red), that is, all sequences that consist of two Bs and two Rs. There are six of them: BBRR, BRBR, BRRB, RBBR, RBRB, and RRBB. They are all mutually exclusive. The probability of each is $26/52 \times 25/51 \times 26/50 \times 25/49$ (not necessarily in that order), which works out to be .0650; .0650 added to itself six times is equal to $6 \times .0650$, or .390—that is, about 4 chances in 10.

An Empirical Demonstration of Probability

Do you understand what is going on? If not, be patient. Probability is tough stuff. Maybe this will help: Shuffle your cards, draw four cards without replacement and record what you drew, using symbols such as AC for the ace of clubs, 7H for the 7 of hearts, and JS for the jack of spades. Repeat the whole process 50 times (i.e., shuffle, draw four cards, record the results), sampling without replacement *within* each drawing of four cards but sampling with replacement *between* each drawing of four cards (do you follow that?). Record the results of your 50 samples before reading on.

Our calculations for Example 4, above, suggest that in approximately 20 of those 50 samples you should get two blacks and two reds. I say "approximately" because (a) .390 added to itself 50 times (i.e., $50 \times .390$) is not exactly 20; (b) you may not give the cards a thorough shuffling each time, which could affect the results; and (c) probability is a "long-run" notion that applies to a conceptually infinite number of trials, and provides no guarantee as to what will happen in a "short-run" set of 50 samples. What I am trying to say is that you may get more than 20 successes (a success being the occurrence of two black cards and two red cards) or fewer than 20 successes. How many *did* you get? I tried it myself and my results are listed in Table 4.2. As you can see, I got 30 successes, 10 more than the expected number, but that can happen "by chance."

Table 4.2 Fifty Samples of Four Cards From the Playing Card Population

Sample Number	Sample	% Black	Sample Number	Sample	% Black
1	10D, 5C, KS, 10S	75	26	KD, 8C, QS, 4H	50*
2	AC, QD, QC, AD	50*	27	KH, 7S, JH, JS	50*
3	KS, 8H, 6S, 10S	75	28	10H, 8H, 10D, JC	25
4	2C, 5D, 4S, AC	75	29	AC, AD, QC, 8D	50*
5	10D, 3S, 2C, 6D	50*	30	4S, 5C, JH, 4D	50*
6	KD, JS, 10D, 7S	50*	31	4S, 2S, AS, KS	100
7	10S, 4S, 8H, 2D	50*	32	8D, 8H, 2C, 5S	50*
8	10C, 6S, 4H, QH	50*	33	2S, 10H, 4D, AH	25
9	QC, 9D, 3S, 3C	75	34	3D, JC, 5H, 3S	50*
10	4S, 9S, 2C, 3H	75	35	3S, 9H, 2S, KD	50*
11	4C, 3C, 9H, KD	50*	36	KC, AH, 8S, 9H	50*
12	QS, 6C, AD, JH	50*	37	6S, JD, 10H, 6C	50*
13	5D, 4C, AH, 3S	50*	38	3S, 10D, 5D, JS	50*
14	10S, 5C, 3D, 2S	75	39	JC, 10C, QH, 7D	50*
15	QC, 6H, JD, 2H	25	40	2C, 2H, 10C, 3D	50*
16	8C, 5H, 9H, 9C	50*	41	8H, JD, 9S, 9C	50*
17	3C, 9C, AC, 4D	75	42	5S, KH, 8C, 5C	75
18	7D, 10C, 8D, 8H	25	43	2S, 3D, 3C, 7H	50*
19	KH, 8H, QS, KC	50*	44	5S, QD, 2S, 8S	75
20	3S, QC, AH, 4S	75	45	KS, 8H, 9C, 5D	50*
21	4H, QC, AH, AC	50*	46	KD, JH, 7S, KS	50*
22	2H, 6C, JH, AD	25	47	AS, 3H, QS, 6D	50*
23	QC, 3C, 10D, 7S	75	48	8D, 7C, 2C, QH	50*
24	QC, 6H, JC, 8C	75	49	10D, 4C, 8C, 5C	75
25	9C, 9S, 5D, 10D	50*	50	4S, JS, 8D, 4C	75

*A "success"—that is, two black cards and two red cards.

Statistics and Sampling Error

Table 4.2 also includes the percentages of black cards in all of my samples. They range from 25 (1 black card out of 4) to 100 (all black cards); "by chance," none of my samples consisted of all red cards (i.e., contained no black cards). The actual percentage of black cards in the population is 50 (the population mean—one of its parameters); 30 of my sample results (sample results are called *statistics*) were exactly equal to that parameter, and the other 20 were not. Whenever a statistic is not equal to its corresponding parameter a *sampling error* has been made. I shall have a great deal to say about sampling errors in Chapter 5.

Let me close this chapter with a few supplementary remarks. First, an assumption that underlies the previous discussion of probability and sampling is that the selection process should be *random*; that is, each of the objects in the population has an equal chance of being selected whenever a sample is drawn. For the population of cards, a thorough shuffling of the deck satisfies the criterion of randomness; in scientific research other devices are employed, such as tables of random numbers. It is essential to understand, however, that it is the *process,* not the outcome, that is random. I might draw 10 cards from a deck with replacement and get the ace of spades every time, but still have a random process. (The probability of getting the ace of spades on each draw is admittedly very small; an extension of Rule 1 gives an answer of 1/52 raised to the tenth power, which is about .00000000000000001, but it *could* happen.)

My second remark concerns the difference between probability and *odds*. The odds against a particular outcome is the ratio of the probability that the outcome will not take place to the probability that it will take place. For example, the odds against drawing a spade in a single draw from a deck of cards is $(3/4)/(1/4) = 3/1$ (which is read as "3 to 1"), not 4/1, as is commonly believed.

Finally, the matter of *sample size.* The question most often asked of statisticians is, What size sample should I employ? The statistician always answers that question with another question: How far wrong can you afford to be when you make your statistical inference? Keep that in mind as you read the next few chapters (indeed, it's a good idea never to forget it). We shall return to this important topic in Chapter 7.

Exercises

1. The first 13 nursery school children are girls, the next 26 are boys, and the last 13 children are girls (see Exercise 1 in Chapter 3). Think of the card color variable as the variable "sex" (1 = black card = female; 0 = red card = male). If you select one child at random, what is the probability that the child will be a girl? What is the probability that you will not select Donna Abbey?

2. If you select two children at random without replacement, what is the probability that they will both be girls?

3. If you select three children at random with replacement, what are the odds against your getting all girls?

4. How many different samples of four states could you draw without replacement from the population of 52 states?

5. For the frequency distribution that you constructed for Exercise 4a in Chapter 1, if you select one state at random, what is the probability that it will have more than 10 members in the House of Representatives?

Sampling Distributions

The basis for all inferences from known sample data to unknown population data is the concept of a *sampling distribution.* If you are content merely to describe the data that you have, you don't need to know anything about sampling distributions, but if you are interested in the problem of generalizing from sample to population, you must understand this concept thoroughly.

Definition of a Sampling Distribution

Let's start with the definition of a sampling distribution, and then take it apart, piece by piece: *A sampling distribution is a frequency distribution of a large number of values of a statistic for samples of the same size randomly drawn from the same population.*

First of all, then, a sampling distribution is a special kind of frequency distribution. You know what a frequency distribution is—we've had lots of those already. The crucial point is: What is it a frequency distribution *of*? That brings us to the second part of the definition. It is a frequency distribution of a statistic. A statistic is a descriptive index for a sample—for example, a sample mean or a sample standard deviation. A sampling distribution is not a distribution of all the observations on a given variable for the population; that would be a *population distribution*. (All of the frequency distributions in Table 1.3 are population distributions.) A sampling distribution is also not a distribution of the observations on a given variable for a sample; that would be a *sample distribution* (see Table 4.1 for an example of a sample distribution).

The same part of the definition tells us that we must have a "large" number of values of a sample mean, a sample standard deviation, or whatever statistic we may be interested in. But how large is large? That depends upon whether you want to talk about a *theoretical sampling distribution*, where you can calculate the frequencies (actually, relative frequencies) of all possible values of the statistic, or an *empirical sampling distribution*, where you must count the frequencies of the values you actually do obtain when drawing repeated samples. In the latter case there is no precise definition of *large*, but 50 values would seem to be a bare minimum. Theoretical sampling distributions are always preferable because they deal with all possible samples and you don't have to worry about the actual mechanics of sampling, but there are some statistics whose theoretical sampling distributions are mathematically indeterminate. For such statistics one has no other choice but to draw sample after sample, manually or by computer, and empirically generate the relevant sampling distributions.

The last part of the definition stipulates that the samples must be of the same size and randomly drawn from the same population. Those conditions are both intuitively reasonable and mathematically necessary. A sampling distribution based on some samples of size 2 and some samples of size 10, or some samples from one population and some samples from another population, would be meaningless and intractable.

Table 5.1 An Empirical Sampling Distribution of % Black for 50 Samples of Size Four Drawn Without Replacement Within Sample and With Replacement Between Samples

Value (of statistic) (%)	Tally	Frequency	Relative Frequency
0	0	.000	
25	11111	5	.100
50	111111111111111111111111111111	30	.600
75	11111111111111	14	.280
100	1	50	.020

NOTE: Mean = 55.5; standard deviation (standard error) = 16.039; skewness = −.788; kurtosis = 3.112.

An Example of an
Empirical Sampling Distribution

I believe it's time for an example. Let's consider the empirical sampling distribution of the percentage of black cards for 50 samples of size four drawn from our playing card population without replacement (within sample); we have to sample with replacement between samples or we would run out of cards after 13 samples. Table 4.2 will serve very nicely to provide the necessary data, because that table includes the percentage of black cards that I obtained in each of my 50 samples. The first value of that statistic is 75, the second is 50, . . . the fiftieth is 75. The desired sampling distribution is a frequency distribution of those 50 values, as displayed in Table 5.1.

That's how you get an empirical sampling distribution. Note some of the surprising and not-so-surprising features of this particular distribution:

1. The value of the true percentage of black cards in the population (the parameter), 50, was obtained 30 times in 50 samples, as pointed out in Chapter 4. It's too bad that I didn't get all 50s, but at least I got more 50s than anything else; that is, 50 is the mode of the sampling distribution.

2. The distribution is skewed to the left (a little heavy on the right). This is intuitively disconcerting because we would expect to be "off on the high side" about as often as we are "off on the low side."

3. Given that 0, 25, 50, 75, and 100 are all of the *possible* values, it is also disconcerting that I didn't get any 0s, but I did get a 100. (I actually drew a second set of 50 samples and got a 0 in one of those samples, but I also got *four* 100s! Do you see what I mean about 50 samples being a bare minimum base for an empirical sampling distribution?)

4. The mean of this sampling distribution is 55.5, the standard deviation is 16.039, the skewness is –.788, and the kurtosis is 3.112. We can get the same sorts of summary indexes for a sampling distribution that we get for any frequency distribution.

5. The relative frequency of each of the sample percentages gives us an approximation of the *probability* of getting the various values when we draw a sample of 4 cards from the population of 52 cards.

The Corresponding
Theoretical Sampling Distribution

Table 5.2 displays the *theoretical* sampling distribution for this same statistic. Let me explain how it was derived:

1. The probability (relative frequency) of 0% (none out of four) black cards is the same as the probability of four red cards (the sequence RRRR), which is equal to $26/52 \times 25/51 \times 24/50 \times 23/49$, or approximately .055.

2. The probability of 25% (one out of four) black cards is the probability of BRRR *or* RBRR *or* RRBR *or* RRRB. Each of these has a probability of $26/52 \times 26/51 \times 25/50 \times 24/49$ (not necessarily in that order), or .0625. Therefore, the probability of 25% black is $4 \times .0625$, or .250.

3. The probability of 50% black, which was calculated in Chapter 4, is .390.

4. Because 75% black = 25% red, and black and red are equally likely, the probability of 75% black is the same as the probability of 25% red, which in turn is the same as the probability of 25% black, the latter of which was already determined to be .250.

5. Similarly, the probability of 100% black = the probability of 0% red = the probability of 0% black = .055.

Table 5.2 The Theoretical Sampling Distribution of % Black for Samples of Size Four Drawn Without Replacement

Value (of statistic) (%)	Relative Frequency
0	.055
25	.250
50	.390
75	.250
100	.055

NOTE: Mean = 50; standard deviation (standard error) = 24.238; skewness = 0; kurtosis = 2.558.

It is of considerable interest to compare the relative frequencies for the empirical sampling distribution with the corresponding theoretical sampling distribution. (I'll bet you did that already, didn't you?) Considering that the empirical distribution is based on only 50 samples, the relative frequencies for the two distributions are not all that different from each other. The fact that they are not identical is no cause for concern, because that is what chance is all about. (If you still have the data for *your* 50 samples—see Chapter 4—construct your own empirical sampling distribution for percentage black and find out if it comes closer to the theoretical sampling distribution than mine did.)

We could extend this example to other statistics, other variables, other sample sizes, other populations, and conditions involving sampling with replacement within sample rather than sampling without replacement. We would get a *different* sampling distribution each time we change any one of these factors. (Remember that sentence and you'll be well on your way toward being an authority on sampling distributions.)

Why Do We Need Sampling Distributions?

We know how to get a sampling distribution, we know how a sampling distribution differs from a population distribution and a sample distribution, and we know there are a lot of sampling distributions. What we don't know yet is why we need them. Until we face up to that, the whole thing is going to seem like a meaningless exercise a statistician might carry out if he or she has nothing better to do.

The reason we need sampling distributions is that most of the time when we carry out scientific studies we will have *one* statistic for *one* sample, and if we don't know how that statistic varies from sample to sample, that is, if we don't know its sampling distribution, we will have no foundation for making any kind of sample-to-population inference.

The matter of "role definition" is important here. The person carrying out the research doesn't actually generate the sampling distribution. He or she has enough to do in choosing the statistic of interest and drawing the sample, to say nothing about formulating the research problem, designing the study, and so on. But *somebody* has to construct sampling distributions so that statistical inferences are possible. Those somebodies are the mathematical statisticians, and the products of their labors are tables and formulas for distributions such as those for the normal, *t*, chi-square, and *F* sampling distributions that are found in the backs of most statistics books (but not this one).

Think of it as a symbiotic process. Some mathematical statistician has to *deduce* the sampling distribution of a statistic for samples of various sizes drawn at random from some population, so that some scientist who has one statistic for one sample can *induce* whether or not that sample came from the specified population. Such an induction is of course always subject to error (because of our old friend chance).

Standard Error

A concept that is very closely associated with sampling distributions is the *standard error*. A standard error is a standard deviation of a sampling distribution. The term *standard error* is actually an abbreviation for *standard deviation of sampling errors* (any statistic that is not equal to the corresponding parameter is a sampling error). Because any standard deviation is a measure of the typical variability around the mean of a frequency distribution, a standard error is a measure of how tightly clustered the statistics are to their own mean (which, for many sampling distributions, is the parameter itself), that is, how much they vary from one another. The larger the sample size, the smaller the standard error, and the more accurate the inference from sample to population is likely to be.

That's all I have to say about sampling distributions for now. If you've got the concept, beautiful. Hang onto it; don't lose it. If you haven't got it, ask for help. We'll keep coming back to it, but in increasingly restrictive contexts. A firm grasp of the general notion is essential at this stage.

In the next two chapters we shall see how sampling distributions are used in the most common kinds of statistical inferences: point estimation, interval estimation, and hypothesis testing.

Exercises

1. If you haven't already done so, take your deck of cards and write on each card the name of the child and the name of the state that corresponds to the card (for example, on the ace of clubs you would write Donna Abbey and Delaware)—see Tables 1.1, 1.4, and 1.5. Then generate an empirical sampling distribution of the percentage of girls for 50 samples of size two sampled without replacement from the nursery school population. (Note that for each sample of size two the only possible values for the statistic "% girls" are 0, 50, and 100.)

2. What do you think would happen to that sampling distribution if you took samples of size five rather than two? Why?

3. What do you think would happen if you took 100 samples rather than 50? Why?

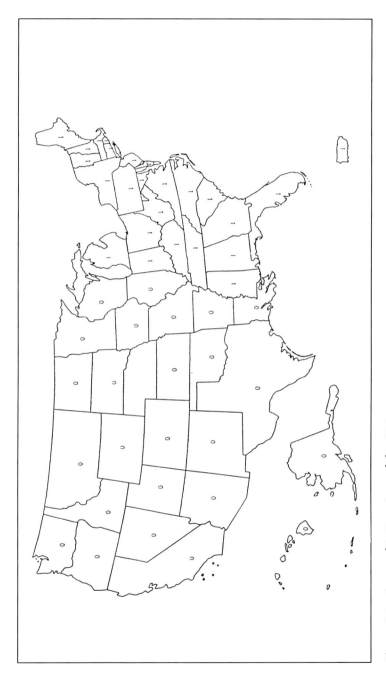

Figure 5.1. Geographic Locations of the 52 States

4. For the 52 states, think of the card color variable as the variable "geographic location" (1 = black = east of the Mississippi River; 0 = red = west of the Mississippi River); choose some sample size, number of samples, and type of sampling (without replacement or with replacement); and generate an empirical sampling distribution of the percentage of east-of-the-Mississippi states in each of the samples. (Use the map in Figure 5.1 and your vast knowledge of geography to determine which states are coded 1 and which states are coded 0. You will note that I have labeled Illinois and Wisconsin as west of the Mississippi River, although both are actually mostly east of the Mississippi. Let's just call this poetic license!)

5. Calculate the standard deviation of the sampling distribution that you generated in the previous exercise, that is, its standard error. How do you interpret that number?

Estimation

There are three kinds of inferences you can make from known sample data to unknown population data. The first kind is called *point estimation* and is a declaration that a particular population parameter is equal to some specified value, based on the value of an obtained sample statistic. For example, you might get a value of 75 for a sample percentage and estimate that the population percentage is also 75.

But if your sample is small, you may not feel comfortable about specifying one value, preferring instead to name two values between which you believe the parameter to lie. For example, you might say that you believe that the population percentage is between 65 and 85. That is an example of *interval estimation*. Both point estimation and interval estimation will be treated in this chapter.

More commonly, however, before you collect any data whatsoever, you make a tentative "guess" that the parameter is equal to some specified value (based on either theory or hunch), and after collecting some sample data, you decide whether your guess was a good one. For example, you might speculate that the population percentage is

60, get a value of 75 for the sample percentage, and reject 60 as a bad guess. That is an example of *hypothesis testing*, which is the topic of Chapter 7.

Point Estimation

Point estimation is the easiest kind of statistical inference to talk about, but it is the kind least often employed, for the reason alluded to above—that is, the smallness of most samples. It's a good place to start, however, because many parameters have in some sense a "best" point estimator. I use the word *estimator* rather than *estimate* for two reasons: (a) It is important to distinguish between what it is we do to the observations in the sample (the estimator) and the number that we arrive at (the estimate); and (b) we are always think-ing in "long-run" terms regarding our optimal strategy. I might make a wild conjecture that a particular parameter is equal to 10 and be lucky enough to be right just that once, but some other well-defined estimation procedure that is "off" that time might be the better bet generally. As I've said before, that's what chance is all about.

It turns out that the sample percentage (the statistic) is the best estimator of the population percentage (the parameter), but it is important to clarify the meaning of *best*. It is best in the sense that it is *unbiased*, which is to say that the mean value of the statistic over repeated samples of the same size is equal to the parameter being estimated; that is, the mean of the sampling distribution of the statistic is equal to the parameter. The sample percentage operates on the sample observations in exactly the same way as the corresponding population percentage operates on the population observations (add up the zeros and ones, divide by the sample size, and multiply by 100), and is therefore an intuitively "best" es-timator. But this does *not* mean that you always hit the population percentage on the button by using the sample percentage to estimate it. For any given sample it might be better to multiply the sample percentage by 1.23 or .59, or whatever. It *does* mean that if you use the sample percentage to estimate the population percentage you will be right "in the long run."

An Example of Point Estimation

But enough of this; let's get to work. Grab your deck of cards and draw a sample of 23 cards, this time *with* replacement (for a little variety); that is, shuffle, draw, record, replace; shuffle, draw, record, replace; and so on, 23 times. I'll do it, too. You make the same calculations for your cards that I do for mine. Here are my cards:

1. 8S
2. 7H
3. AC
4. QC
5. 2C
6. QS
7. 4D
8. 3D
9. 2D
10. 7C
11. 10D
12. 6S
13. JH
14. 9S
15. 9C
16. 10C
17. 4C
18. 4H
19. JD
20. 3C
21. 10C (a repeat)
22. 9C (a repeat)
23. QC (a repeat)

Note that I got three repeats, but no re-repeats; that is, I did not draw any given card three or more times. How many of each did you get?

For the color variable, I have the following observations for these 23 cards (0 = red; 1 = black):

1. 1
2. 0
3. 1
4. 1
5. 1
6. 1
7. 0
8. 0
9. 0
10. 1
11. 0
12. 1
13. 0
14. 1
15. 1
16. 1
17. 1
18. 0
19. 0
20. 1
21. 1
22. 1
23. 1

Because there are 15 1s (black cards) and 8 0s (red cards), the percentage of black cards in my sample is (15/23) × 100, or about 65.2. What should I infer about the population percentage? In Chapter 2 we convinced ourselves that a percentage is a special kind of mean, so my unbiased estimate of the population mean (the population percentage) is 65.2. That's too bad (the true value is actually 50), but it happens.

There's not much else I can say about point estimation. Besides, I'm anxious to get on to interval estimation (my favorite kind of statistical inference), so let's do that.

Interval Estimation

Because it is usually presumptuous to infer a single value for an unknown population parameter on the basis of a small sample, a more defensible procedure is to specify a range of values within which the parameter is alleged to lie. The procedure is called interval estimation (as opposed to point estimation) and the resulting set of values is called a *confidence interval.* The person making the inference has some specified amount of confidence that the obtained interval "captures" the relevant parameter.

The key to interval estimation is the concept of a standard error (which we treated at the end of Chapter 5) in conjunction with the shape of the sampling distribution of the statistic employed in the estimation process. As I pointed out in Chapter 5, the standard error of a statistic is the standard deviation of its sampling distribution, and is a measure of how much a statistic for one sample of a given size tends to vary from a statistic for another sample of the same size from the same population. If the sampling distribution is of the "normal" (bell-shaped) form, it can be shown that the probability is about .68 that a statistic will lie within one standard error of its corresponding parameter, is about .95 that it will lie within two standard errors, and so on. By turning this argument inside out, so to speak, if you have one sample statistic (which is usually the case) and you lay off one standard error to the left and one standard error to the right, you can say you have a confidence level of .68 that the parameter is captured in your interval; if you lay off two standard errors left and right, you can say you have a confidence level of .95 (i.e., you can be more confident) that the parameter is captured; and so on. The greater confidence you desire, the wider the interval must be (all other things being equal).

It is important to understand that this "tie-in" between one standard error and .68 confidence, between two standard errors and

.95 confidence, and so on, holds only for normal sampling distributions. If the sampling distribution for a particular statistic of interest is *not* normal (and many of them are not), you may have to lay off either fewer than two standard errors or more than two standard errors for .95 confidence, for example. If the shape of the sampling distribution is unknown and indeterminate, you really don't know how many standard errors to lay off for various confidence levels. Do you see now why sampling distributions are so important?

The sample percentage is a statistic that does have an approximately normal sampling distribution, especially if we take a large sample with replacement. If the population distribution is approximately symmetric—that is, if the population percentage is not too far from 50—and if the sample size is not too small (about 20 or more), we can get a .95 confidence interval for the population percentage as follows:

1. Find the sample percentage (the statistic).
2. Calculate the standard error of the sample percentage. The mathematical statisticians tell us that the standard error of the sample percentage is approximately equal to the square root of the quantity obtained by multiplying the population percentage by the difference between the population percentage and 100 and dividing by the sample size, if the sampling has been with replacement. (If the sampling has been without replacement, this must be further multiplied by "the finite population correction factor," which is the square root of the quotient of the population size minus the sample size and the population size minus one.) This presents a bit of a dilemma because the population percentage is unknown (that's what we're trying to estimate), so we have to plug in the sample percentage instead. That sounds a little strange, but there's nothing else we can do, and fortunately the product of a percentage times the difference between a percentage and 100 is very close to 2,500 for percentages that are very close to 50. (*Special note:* In a couple of the exercises at the end of Chapter 5, you calculated the standard error of your empirical sampling distributions the same way a standard deviation is usually calculated, but you had just one set of samples, not all possible samples. The procedure just described applies to the *theoretical* sampling distribution for sample percentages.)

3. Lay off two standard errors to the left of the sample percentage and two standard errors to the right of the sample percentage. (It's really 1.96 standard errors, but 2 is close enough for our purposes.) The interval thus established is said to have a probability of .95 of capturing (including, or "bracketing") the population percentage.

An Example of Interval Estimation

Let's work through an example, using the 23 sample observations for the color variable displayed above:

1. The percentage of black cards in my sample is 65.2.
2. My estimate of the standard error is the square root of the product of 65.2 and 34.8 divided by 23, or 9.9. (Did you follow that? Be sure to check all of these calculations, slowly and carefully.)
3. My .95 confidence interval for the population percentage therefore extends from $65.2 - 2(9.9)$ to $65.2 + 2(9.9)$—that is, from 45.4 to 85.0. The probability is .95 that the interval from 45.4% to 85.0% brackets the population value. The probability is .05 that it does not. As the true population percentage is 50, my inference is correct this time (again, in real life I wouldn't know that), but it won't always be.

So if I had to give one number that is my best single (point) estimate of the percentage black, based on this sample of 23 observations, I would say 65.2, but I would have little or no confidence in that estimate. If I could give an estimate of an interval within which I believe that parameter to lie, I would say it was from 45.4 to 85.0. That wouldn't narrow things down very much (the sample size is a bit small), but I would have a probability of .95 of making a correct inference.

There is of course nothing special about a .95 confidence interval (other than the fact that it is conventional). The procedure is exactly the same for the .68, .99, or any of the other popular confidence levels. All that will change is the number of standard errors that you lay off (one standard error for .68; 2.58 standard errors for .99;

and so on—see any table of the normal distribution for the necessary values). If you want to be very confident that you have captured the parameter, you must give yourself a lot of leeway—that is, lay off a lot of standard errors.

Get the picture? Why don't you take the sample of 23 cards that you drew, write down the color of each, and construct your .95 confidence interval for the percentage black in the population?

Exercises

1. Draw a random sample of 30 nursery school children, with replacement, and calculate an unbiased estimate (point estimate) of the percentage of girls in the population.

2. Draw a second sample of the same size in the same way and get another estimate of that same parameter. Then combine the two samples. Which estimate is closest to the true population percentage (50)—the estimate based on the first sample of 30 girls, the estimate based on the second sample of 30 girls, the mean of those two estimates, or the estimate based on the combined sample of 60 girls? Does your result make sense? Why or why not?

3. Use your first sample of 30 to construct a .95 confidence interval for the percentage of girls in the population. Did your interval include what you know (but wouldn't know in real life) to be the true parameter?

4. Use your second sample of 30 but switch variables to construct a .95 confidence interval for the percentage of states east of the Mississippi River. Did you "win" or "lose"? Would a .68 confidence interval be wider or narrower than your .95 confidence interval? Why? Would a .95 confidence interval for a sample of 60 be wider or narrower than a .95 confidence interval for a sample of 30? Why?

5. For the combined sample of 60 states, construct a .95 confidence interval for the percentage of states with fewer than 10 representatives and phrase the appropriate inference. Was that a fairly wide or a fairly narrow interval? Were you surprised? Why or why not?

Hypothesis Testing

We come now to the most popular method of statistical inference. I would venture to say that hypothesis testing procedures are used in at least 90% of all research in which sample-to-population inferences are made. I think that is unfortunate because, as we shall see, hypothesis testing is a rather awkward way to approach the inference problem and should be confined to a fairly small subset of applications where the actual magnitude of a parameter is of no interest, but its equality or nonequality to some specified value is.

One of the bothersome things about hypothesis testing is all of the jargon that is associated with it. The research literature is sprinkled with such terms, however, so you'd better get used to them. Let's take the percentage black problem discussed in Chapter 6, put it in the hypothesis testing framework, and make the warranted inference.

A Previous Example Reconsidered

You remember the situation. We have decided to take a random sample of 23 cards drawn with replacement and we are interested in the percentage of black cards (the parameter) in the population from which the sample is to be drawn. The first step in hypothesis testing is to state a hypothesis (that sounds reasonable) regarding the parameter, *before we draw the sample*. (It would be cheating, wouldn't it, to state a hypothesis *after* we see some data?) But what hypothesis?

Null and Alternative Hypotheses

The hypothesis that is always tested is something called a *null hypothesis*. It is called a null hypothesis for a variety of reasons: (a) The hypothesized value for the parameter is often zero; (b) it is the "conservative," "nothing special is going on" hypothesis; and (c) the researcher usually hopes that the sample data will "nullify"— that is, reject—that hypothesis.

For our example, the null hypothesis that would be put to test is as follows: The percentage black is equal to 50. Although 50 is not zero, that hypothesis is the "nothing special is going on" hypothesis, because ordinary playing card populations have 50% black cards (and 50% red cards), and *if* something special should be going on—that is, if we have sampled an unusual deck of cards—then we would want to be able to reject the hypothesis that we have a usual deck. (Do you follow that? I told you that hypothesis testing is strange.)

Things are actually even a bit more complicated. You have to test *two* hypotheses *against* each other: the "conservative" null hypothesis, which is one guess about a population parameter, and a "liberal" *alternative hypothesis*, which is another guess about the same parameter. For our example, the alternative hypothesis might be any one of the following:

- *Alternative Hypothesis 1:* The percentage black is not equal to 50. (This is the simple denial of the null hypothesis.)
- *Alternative Hypothesis 2:* The percentage black is greater than 50. (This would be the appropriate hypothesis if our theory or hunch were that the deck we'll be sampling has an unusually high percentage of black cards.)
- *Alternative Hypothesis 3:* The percentage black is less than 50. (This would be the appropriate alternative hypothesis if our theory or hunch were that the deck we'll be sampling has an unusually low percentage of black cards.)
- *Alternative Hypothesis 4:* The percentage black is equal to 60 (or 40, or 81, or whatever our theory or hunch might be).

The first of these is nonspecific and nondirectional, as it does not postulate any particular value for the parameter and doesn't even stipulate whether the parameter is greater or less than the value hypothesized in the null. The second and third alternatives are also nonspecific, but they are directional, the former claiming that the parameter is greater than 50 and the latter claiming that the parameter is less than 50. The fourth alternative is both specific and directional; a particular value is hypothesized, and, being specific, it must be on one side or the other of the value hypothesized in the null.

Back to the Example

Let's say that we want to test 50 against not-50. We draw our sample and get 65.2% black cards. Because 65.2 is not 50, the null hypothesis should automatically be declared false and the alternative hypothesis should automatically be declared true, right? Wrong, for the following reasons:

1. Our hypotheses are concerned with the *population,* not with the sample.
2. Although we got 65.2% black in the sample, there could be 50% black in the population and our sample result merely a fluke—that is, a sampling error. Keep in mind the distinction between a parameter and a statistic.

This is not to say that after considerable thought, and a few calculations, we won't decide to reject the null hypothesis after all (how's that for a quadruple negative?), but we must not (another negative) be too hasty.

Here's what we have to do. We have to determine the *probability* of getting a difference of 15.2% (= 65.2% – 50%) or more black cards in a sample of 23 cards *if* the population percentage is 50. If that probability is low (less than .05, say), then we would have sufficient evidence for rejecting the null hypothesis in favor of the alternative. If that probability is high (greater than or equal to .05, for example), then the evidence would not be sufficient to reject the null hypothesis. (Do you follow that? If so, great. If not, hang on; it will come to you.)

We find that probability by using the sampling distribution for percentage black, and its standard error (just as we did in the previous chapter for the interval estimation approach to statistical inference), as follows:

1. Because the sampling distribution of percentage black in the sample is normal (if the percentage black in the population is close to 50 and the sample size is not too small), we can determine the probability that *any* sample percentage will differ from the population percentage in terms of numbers of standard errors.

2. The standard error of a sample percentage is the square root of the product of the population percentage and 100 minus the population percentage divided by the sample size, so *if* the population percentage is 50 the standard error is approximately equal to the square root of $50 \times 50 / 23$ or 10.4. This number differs slightly from the 9.9 obtained in Chapter 6 because there we used 65.2 rather than 50 to calculate the standard error. We don't do that here. We can (nay, must) use the value of the parameter stipulated in the null hypothesis that we're testing.

3. Our sample percentage of 65.2 differs from 50 by 15.2 percentage points. Because the standard error is 10.4 points, the discrepancy between our obtained statistic and the hypothesized parameter is $15.2/10.4 = 1.46$ standard errors. Therefore, the probability of getting a discrepancy of 15.2% or more is the probability that any measurement in a normal distribution will differ from its mean by more than 1.46 standard devia-

tions. We could look that up in a table of the normal distribution, but we know that the probability must be greater than .05, because the discrepancy would have to be two or more standard errors for the probability to be that small. Therefore, we do not have sufficient evidence to reject the null hypothesis; there is a reasonably large probability that our sample has been drawn from a population in which the percentage black is 50.

That's the way hypothesis testing always works. You formulate two hypotheses regarding some parameter (one null and one alternative); you draw a sample; you calculate the corresponding statistic; you use the sampling distribution of that statistic to determine the probability of getting a difference between statistic and parameter equal to or greater than the one you got; and you reject or fail to reject the null hypothesis according to whether that probability is small or large.

Your Data

In order to get a feel right now for hypothesis testing, take the data for your sample of 23 cards and go through the same steps I did, for the same null (the parameter is 50) and the same alternative (the parameter is not 50). What was your decision regarding the null?

Type I and Type II Errors

I was lucky. I didn't reject the null, and the null was true. (In real life I wouldn't know whether the null was true or false.) If I had rejected a true null I would have made a mistake. Such a mistake is called a *Type I error*. But there is another kind of mistake I could have made. I could actually have been sampling a "phony" deck of cards that didn't have an equal number of black and red cards, in which case I would have not rejected a false null hypothesis. That kind of mistake is called, naturally enough, a *Type II error*. How about your inference? Did you make an error? If so, was it Type I or Type II?

One- and Two-Tailed Tests

The test of a null hypothesis against its simple denial is called a *two-tailed test* because it involves both ends (tails) of the sampling distribution (discrepancies "on the high side" as well as discrepancies "on the low side"). If we want to test a null hypothesis against certain other kinds of alternative hypotheses (directional/specific or directional/nonspecific), we must use a *one-tailed test* that involves discrepancies on *either* the high side *or* the low side, but not both.

Level of Significance

The probability that we regard as "small" (e.g., the .05 referred to above) is called the *level of significance* or *significance level,* and if the probability of a particular outcome is less than that value the null hypothesis is rejected and the outcome is said to be *statistically significant* (my difference of 15.2% was not statistically significant). The level of significance is therefore the probability of making a Type I error. It is "researcher's choice" as to what level of significance should be used (the choice should depend upon the consequences of making a Type I error), but .05, .01, and .001 are the popular ones.

Power

Determining the probability of making a Type II error is much more complicated. It depends upon what the alternative to the null is. If the value of the parameter postulated in the alternative hypothesis is very close to the value postulated in the null, the probability of making a Type II error is high (unless the sample size is very large), because the obtained statistic will be commensurate with either hypothesized value; so if the alternative is true—that is, the null is false—the researcher will have a high probability of "sticking with" the null when it is false. On the other hand, if the two hypothesized values are not very close to each other, the probability of making a Type II error is low, because the obtained statistic will not be commensurate with both of them; so if the null is false, the researcher will have a low probability of sticking with it. (Do you follow that? This point is *really* important.)

The probability of making a Type II error does not have a special name (comparable to level of significance for Type I error), but its complement, the probability of *not* making a Type II error (i.e., one minus that probability) does. It's called the *power* of the test (the test of the null against the alternative). Because we want to have a low probability of making a Type II error, we want to have a high probability of not making a Type II error; that is, we want a "high-powered" test. Power is a function of sample size; the larger the sample size, the greater the power (all other things being equal).

Testing the Difference
Between Two Percentages

The test of a null hypothesis regarding a particular value of a population percentage is fairly common in survey research. But an application of hypothesis testing that permeates just about all kinds of research is the test of the *difference* between *two* population percentages—for example, the percentage of smokers who get lung cancer and the percentage of nonsmokers who get lung cancer. The null hypothesis in all such applications is that the difference is equal to zero; the alternative hypothesis is usually that the difference is not equal to zero, but occasionally some particular value such as 10% or 20% will be stipulated.

Why the interest in zero versus nonzero? A zero difference would indicate that nothing special is going on; a nonzero difference would suggest that something special *is* going on. For the smoking/lung cancer example, if there is a difference between smokers and non-smokers it would not only be interesting, but perhaps something could be done about it (a special educational effort directed at smokers by the medical community, perhaps).

The test proceeds as follows:

1. The null and alternative hypotheses are stated.

2. A sample is drawn at random from one of the populations and another sample, independent of the first sample, is drawn from the other population. (The two samples are independent whenever they are not "matched" in any way. There are some advantages and some disadvantages to using independent samples—for example, you don't need to worry about what to match the two samples *on*—but if you use matched samples and you're smart enough to have matched the samples on the "right" variable[s], you have a better test.)

3. The percentage black (or female or Catholic or whatever) for each sample and the difference between the two percentages are calculated.

4. The standard error of the difference between two independent percentages is calculated. The mathematical statisticians tell us that the standard error of that statistic is found by taking the square root of the following triple product: the percentage of 1s in the two samples combined times 100 minus that percentage times the sum of the reciprocals of the sample sizes. (That's a mess, isn't it? But hang on; I'll go through all of the calculations in a second.)

5. Divide the difference between the two sample percentages by the standard error, refer that quotient to the normal sampling distribution, and reject or don't reject the null hypothesis (of no difference in the population percentages) accordingly.

An Example

Now for an example. I'll use the same 23 cards I drew before as one of the samples, and I've drawn 21 cards from a different deck of cards to provide the data for a second sample. Table 7.1 contains the cards that constitute each of the samples, the observations for the color variable, the percentage of black cards in each sample, the difference between those two percentages, and the standard error of the difference. Let's use these data to illustrate the procedure for testing the significance of the difference between two independent sample percentages.

1. Null hypothesis: The difference between the two population percentages is equal to zero.

Table 7.1   An Example of a Test of the Significance of the Difference
Between Two Independent Sample Percentages

Card	Observation	Card	Observation	Card	Observation
Sample 1 (23 observations)					
8S	1	10D	0	10C	1
7H	0	6S	1	9C	1
AC	1	JH	0	QC	1
QC	1	9S	1	% black = 65.2	
2C	1	9C	1		
QS	1	10C	1		
4D	0	4C	1		
3D	0	4H	0		
2D	0	JD	0		
7C	1	3C	1		
Sample 2 (21 observations)[a]					
4S	1	7S	1	AS	1
8H	0	3C	1	% black = 66.7	
3C	1	AS	1	difference = 66.7 − 65.2 = 1.5	
KD	0	KC	1	standard error = 14.3[b]	
AH	0	3C	1		
9S	1	8C	1		
KH	0	4H	0		
JC	1	9D	0		
4H	0	3S	1		
10C	1	AS	1		

a. This sample had several repeats, re-repeats, and re-re-repeats, including the ace of spades twice
in succession.
b. The difference is less than one standard error (not statistically significant). The null hypothesis
is "accepted" (not rejected).

Alternative hypothesis (one of several possibilities): The difference between the two population percentages is equal to 20.

2. I've got my two samples. I drew them independently, from two
different populations (decks of cards)—one from each population. My two sample sizes are not equal. They don't have to
be, but power (see above) is maximized when they are similar.

3. The percentage of black cards in my first sample is 65.2; the
percentage of black cards in my second sample is 66.7. The
difference is 1.5.

4. The number of 1s in the combined sample is 15 + 14, or 29. The *percentage* of 1s in the combined sample is $(29/44) \times 100$, or 65.9. The standard error is therefore the square root of the expression $65.9 \times 34.1 \times (1/23 + 1/21)$, which is equal to 14.3.

5. The difference between the two sample percentages, 1.5, is less than one standard error, so the probability of getting such a difference, if the null hypothesis is true, is much greater than .05. Therefore the null cannot be rejected and the difference of 1.5% is not statistically significant.

A few remarks are in order here. First, because the difference between the two *sample* percentages is 1.5, which is much closer to 0 than to 20, the evidence clearly supports the null hypothesis.

Second, and closely related to the first remark, I may have just made a Type II error. That is, the null hypothesis could be false (I know it's true because both of my decks have 50% black cards, but in real life I wouldn't know that) and the alternative hypothesis (that there is a 20-point difference in percentage black) could be true, but my sample sizes are just too small for me to have been able to make the correct inference. In our recently acquired statistical jargon, I may not have had enough power. The probability that I have made a Type II error can actually be calculated for this example, but it's a bit complicated, so I won't bore you with the calculations (the answer is about .70, which is a very high error probability). The probability that I have made a *Type I* error—that is, that I have rejected a true null hypothesis—is actually equal to zero (no matter what significance level I may have implicitly been using), because I didn't reject it. *Before* you make your inference you have some nonzero probabilities of making both kinds of errors, but *after* you make your inference you have only one kind of error to worry about. That's sort of comforting, isn't it?

My third remark concerns the matter of combining the data for the two samples to get a single estimate of the percentage black. *If* the null hypothesis is true—that is, if both populations have the same percentage black—you can get a better estimate of what that common percentage is by "pooling" the data for the two samples than you can get for either of them, because the pooled estimate is based on 44 rather than 23 or 21 observations.

Table 7.2 Approximate Sample Sizes for an Optimal Test of the Statistical Significance of the Difference Between Two Equal-Sized Independent Sample Percentages

Difference Specified in the Alternative Hypothesis (%)	Significance Level					
	.05			.01		
	Desired Power			Desired Power		
	.80	.95	.99	.80	.95	.99
10	392	650	919	584	891	1202
20	98	162	230	146	223	300
30	44	72	102	65	99	134
40	25	41	57	36	56	75
50	13	21	30	19	29	40

SOURCE: Figures are from Tables 6.2.1 and 6.4.1 in Cohen, J. (1988). *Statistical power analysis for the behavioral sciences* (2nd ed.). Hillsdale, NJ: Lawrence Erlbaum.
NOTE: Two-tailed test; table values are the number of observations in *each* sample.

Power and Sample Size

I would like to close this chapter by pursuing the matter of power and its relationship to sample size. There is obviously nothing special about having 23 and 21 observations in the two samples. How many *should* I have drawn? Do you remember what I said in a previous chapter about the question most often asked of statisticians, and what their reply is? How many observations you should have in each sample depends upon how far wrong you can afford to be when you make your inference. In fact, it depends on three things:

1. the alternative hypothesis you're testing against the null;
2. your chosen significance level—that is, the risk you're willing to take of rejecting a true null; in other words, the probability of making a Type I error (*before* seeing the sample data); and
3. the power you desire—that is, the probability of *not* making a Type II error (again, *before* seeing the sample data).

Formulas and tables for selecting sample sizes are provided in many statistics textbooks. I have constructed an abbreviated table (Table 7.2) of sample sizes that are recommended for testing the significance of the difference between two independent sample percentages for typical significance levels and desired powers. As you can see from that table, in order to test the null hypothesis of no difference against an alternative hypothesis of a 20% difference, if I wanted to have "equal protection" against Type I error and Type II error of .05 (power = .95), I should have drawn 162 observations from each of my populations, not 23 or 21. For those small sample sizes the probability is much less than .80 that I would reject the null if it were false. With sample sizes around 25, the difference in the two population percentages would have to be 40 or more for me to have even a .80 probability of rejecting a false null hypothesis (because 25 is the appropriate sample size for a difference of 40%, the .05 significance level, and .80 power).

As you can see, very large sample sizes are required to test for small percentage differences. That makes sense, because it is hard to differentiate between a null hypothesis of no difference and an alternative hypothesis of a little difference. (I didn't even include the sample sizes required for testing differences such as 1% or 2%, but believe me, they are astronomical.)

Exercises

1. Draw a random sample of 25 nursery school children, with replacement, and test the null hypothesis that girls and boys are equally represented in that population (i.e., that the percentage of girls = 50) against the alternative hypothesis that they are not. Did you make the correct or the incorrect inference? If you made the incorrect inference, what kind of error did you make—Type I or Type II? What are some reasons you might have been "forced" to make the incorrect inference?

2. Draw another random sample of 25 children (also with replacement) and test the same null hypothesis again. Was your decision the same or different? Were you surprised? Why or why not?

3. Using these two samples, test the null hypothesis that the two sampled populations have the same percentage of girls. Use the nonspecific, nondirectional alternative hypothesis that the two populations do not have the same percentage of girls. Did you make the correct or the incorrect inference? If you made the correct inference in Exercise 1 *and* in Exercise 2, would you have to make the correct inference here? Why or why not?

4. Dichotomize the 52 states into two subpopulations: fewer than five representatives versus five or more representatives. Draw a random sample of 30 states, with replacement, from each of those subpopulations and test the null hypothesis that the two subpopulations have equal percentages of states that are east of the Mississippi River (against the alternative hypothesis that they do not).

5. Is the null hypothesis in Exercise 4 true or false? Did you reject it? Was your inference correct or incorrect? Was the sample size of 30 appropriate? Why or why not?

2 × 2 Contingency Tables

Information regarding the difference between two independent percentages is often displayed in a *contingency table* (sometimes called a *cross-tabulation* or *cross-tab*). Contingency tables are also useful in conjunction with a technique called *elaboration*. This chapter is devoted to such matters.

Displaying Frequency
Data in a 2 × 2 Table

Let's start with an example. I'll use the same example I exploited near the end of Chapter 7, that is, the test of the difference between the percentage of black cards in one deck of cards and the percentage of black cards in another deck of cards (see Table 7.1 for the raw data). Here is the way the principal information is often displayed:

	Deck 2	_Deck 1_	
Black	14 (66.7%)	15 (65.2%)	29
Non-black	7 (33.3%)	8 (34.8%)	15
(Red)	21	23	44

This is called a 2×2 (two-by-two) contingency table because it has two rows (horizontal) and two columns (vertical)—the other numbers, 29, 15, 21, 23, and 44, are "marginal" totals. The convention almost universally followed is to use as column headings the categories of the _independent_ variable—the potential "cause" (in this case, the type of deck)—and to use as row headings the categories of the _dependent_ variable—the potential "effect" (in this case, the color of the card).

Doing the Percentaging
and Comparing the Percentages

The "percentaging" is done by columns (we take the 15 out of the 23 and get 65.2%, for example, _not_ out of the 29 and _not_ out of the 44) and the resulting percentages are compared across the rows (for example, the 65.2 against the 66.7, just as we did in Chapter 7).

Several cautions must be observed. First of all, the percentages must total 100 for each of the columns, as explained in Chapter 3. Second, the _observations_ as well as the samples must be independent of one another. This is a complex topic, but the thing that most often produces non-independent observations is counting a particular object in more than one category. Finally, the total sample size should be reasonably large, as pointed out in Chapter 7.

Relative Risks and Odds Ratios

Although the emphasis is usually placed on the _difference_ between the two percentages in the first row of the contingency table, it is fairly common in certain research studies, primarily in epidemiology, to emphasize the _quotient_ of those percentages in addition to, or instead of, their difference. That quotient is often referred to as the *relative risk* (for reasons associated with the jargon of epidemiological research). For our example the quotient of 66.7 and 65.2 is 1.02, so the relative risk of Deck 2's yielding a black card—compared with Deck 1—is 1.02, that is, 2% higher for Deck 2 than for Deck 1. (Some researchers prefer to put the smaller percentage in the numerator and the larger percentage in the denominator. For our example the relative risk of Deck 1's yielding a black card would be 65.2/66.7, or .98.)

There is another concept associated with relative risk that is of even greater interest to epidemiologists: the *odds ratio.* It is computed by dividing the product of the upper-left corner frequency in the 2 × 2 table and the lower-right frequency by the product of the upper-right and lower-left frequencies. For our table that ratio is (14 × 8)/(15 × 7) = 1.07. The odds ratio is a good approximation to the relative risk when the two percentages being compared are very close to one another and when the relative frequency of the "disease" is small. (In our example the "disease" is "yielding a black card.") The odds ratio and its logarithm have very nice mathematical properties.

Elaboration

There are occasions when we would like to explore the data further by statistically controlling for one or more variables that might affect the simple difference between two percentages. For example, in cigarette smoking/lung cancer research the investigator might not be content merely to compare the difference in percentage of lung cancer for smokers versus non-smokers. It could be suspected that smokers are more likely to live in areas that have a great deal of air pollution and that non-smokers are more likely to live in areas that have little or no air pollution. It would therefore be of considerable interest to see if the difference in percentage of lung cancer for smokers and non-smokers is approximately the same for people living in heavily polluted areas as for people living in lightly polluted areas. This would necessitate the construction of *two* 2×2 tables, one for heavy pollution and one for light pollution.

We can illustrate this by using our deck of cards example. Think of Deck 2 as Smokers and Deck 1 as Non-smokers, black card as Lung Cancer and nonblack card as No Lung Cancer, and face card as Heavy Pollution and non-face card as Light Pollution. Referring to the actual cards drawn (see Table 7.1), the required 2×2 tables are as follows (be sure that you check my numbers). For face cards:

	Deck 2	Deck 1	
Black	2 (50%)	3 (60%)	5
Non-black	2 (50%)	2 (40%)	4
	4	5	9

For non-face cards:

	Deck 2	Deck 1	
Black	12 (70.6%)	12 (66.7%)	24
Non-black	5 (29.4%)	6 (33.3%)	11
	17	18	35

Although the actual frequencies are very small, from these elaborated tables we can see that when controlling for pictureness, the difference in percentage black is 60% − 50% = 10% "in favor of" Deck 1 for face cards and 70.6% − 66.7% = 3.9% "in favor of" Deck 2 for non-face cards (versus 1.5% "overall"). Therefore, the results are fairly similar whether or not pictureness is controlled. Imposing the smoking/lung cancer vocabulary on this artificial example, we would say that the effect of cigarette smoking on lung cancer is essentially the same in heavily polluted areas as it is in lightly polluted areas.

Dependent Samples

So far it has been taken for granted that the two samples being compared are independent of one another, i.e., that the *samples* as well as the *observations* are independent. (The words "independent" and "dependent" come up in a number of contexts, so you'd better get used to that!) Sometimes when we're interested in testing the difference between two percentages the samples are "matched" with one another and are therefore "dependent" rather than independent. The matching can come about in a number of ways. The two samples may consist of the same people being measured on two separate occasions; or pairs of people (e.g., identical twins) being measured once each; or pairs of people being randomly assigned to experimental treatments, with one member of each pair assigned to the experimental group and the other member assigned to the control group, and all measured on the dependent variable at the end of the experiment. Whenever matching is a feature of the design, it must be taken into account in the analysis.

Use your imigination and suppose that Deck 2 in the above example had been manufactured at the same playing card company (there are such places) as Deck1. In addition, suppose that the observations from the two decks had been "matched" by shuffling both decks the same number of times, drawing a sample of 22 cards from each deck (with replacement) by pairing the first card from Deck 2 with first card from deck1; the second card from Deck 2 with the second card from Deck 1; and so on. This produces the following matched pairs of the data (the data are the same as in Table 7.1, with one observation "stolen" from Deck 1 and "given" to Deck 2):

Pair	Deck 2 Obsevation	Deck1 Obsevation
1	1	1
2	0	0
3	1	1
4	0	1
5	0	1
6	1	1
7	0	0
8	1	0
9	0	0

Pair	Deck 2 Obsevation	Deck 1 Obsevation
10	1	1
11	1	0
12	1	1
13	1	0
14	1	1
15	1	1
16	1	1
17	0	1
18	0	0
19	1	0
20	1	1
21	1	1
22	1	1

The percentages of 1s in Sample 2 (from Deck 2) is 15 divided by 22 times 100, or 68.2%. The percentages of 1s in Sample 1 (from Deck 1) is 14 divided by 22 times 100, or 63.6%. The actual frequencies, and corresponding percentages, can be displayed in the following 2 × 2 table:

		Sample 1		
		0	1	
Sample 2	1	4	11	15 (68.2%)
	0	4	3	
			14 (63.6%)	

Note the difference in the labeling of the rows and columns for a 2 × 2 tabel for dependent samples versus a 2 × 2 table for independent samples.

In order to test the significance of the difference between these two percentages (i.e., 68.2% – 63.6% = 4.6%), you take the difference between the frequency in the upper-left corner of the 2 × 2 table and the frequency in the lower-right corner, divide that by the square root of the sum of those two frequencies, and compare that quotient to our old normal-curve friends 1.96 (for the .05 level of significance), 2.58 (for .01), and so on. For this example we get 4 – 3, divided by the square root of 4 + 3, which is 1/2.646 or .378. Because this is much less than 1.96, the difference between the two percentages is not statistically significant. This is not surprising because the difference is only 4.6% and the sample sizes are fairly small, so the "matching" didn't seem to help much. But for the real-world research on identical twins, people tested before and after some event of considerable interest, the dependent-sample approach often pays high dividends. (Interest; dividends—get it?)

Partially Independent, Partially Dependent Samples

Believe it or not, there are occasions on which you might have two samples that are "partially matched." For example, in studying the attitudes of men and women toward abortion you might have two samples, one consisting of wives and the other consisting of husbands, but you may not have all actual wife-husband pairs. (One or more of the spouses might refuse to participate in the study, or whatever.) As you might expect, the analysis of the data gets very complicated in this case. If this should happen to you, please consult your local statistician!

Exercises

1. Create two decks (populations) of cards out of your single deck. Let the first deck (population) consist of all of the clubs, all of the diamonds, and the heart face cards. Let the second deck (population) consist of the rest of the hearts and all of the spades. Make a frequency distribution of the sex variable for each of the two artificial populations of nursery school children, remembering that black = female and red = male. (There should be 13/29 = 44.8% girls in the first population and 13/23 = 56.5% girls in the second population.) Draw a sample of 20 cards, *with replacement,* from each of the two populations and determine the percentage of girls in each of the two samples and the difference between the two sample percentages. Test the null hypothesis that those two samples come from populations that have the same percentage of females, using the procedure outlined in Chapter 7. (You know that this hypothesis is false, because you have created two populations having different percentages of females, but . . . . Complete this sentence in 25 words or less.) Was your decision regarding the null hypothesis correct or incorrect? If incorrect, what kind of error did you make, Type I or Type II? Why?

2. Display the sample data for Exercise 1 in a 2 × 2 contingency table, with all of the sample frequencies and associated percentages. Calculate the relative risk of yielding girls for the two samples, and also the sample odds ratio. Are those two numbers fairly close or not? Why do you think that is?

3. Now suppose that you had taken a sample of 200, rather than 20, from each of the two populations, but the percentages of girls in the two samples were the same as for the two samples of 20. What do you think your decision regarding the null hypothesis would have been? Would you be more likely, less likely, or equally likely to make a Type I error? A Type II error? What effect, if any, would that have on the relative risk and the odds ratio? Why?

4. Recasting these populations in terms of the 52 states, with the first 29 states constituting Population 1 and the last 23 making up Population 2, interpret Exercises 1-3 for the east-of-the-Mississippi versus west-of-the-Mississippi dichotomy.

5. Set up the two 2×2 contingency tables that you would use to test the difference between east and west percentages for those states that have fewer than 10 representatives and for those that have 10 or more representatives. Did things change much when you controlled for number of representatives? Why do you think that may be?

Where Do You Go From Here?

There is much more to the study of statistics than I have been able to cram into the preceding eight chapters, but I assure you that we have covered the essential concepts, all of which can be subsumed under the following key terms:

- Population
- Parameter
- Sample
- Statistic
- Sampling distribution

Sometimes you have access to the entire population of interest, in which case you make your measurements and calculate the relevant parameter(s). Most of the time you don't, so you take a sample from the population, calculate a statistic for that sample, and, by using the appropriate sampling distribution, either estimate or test a hypothesis about the corresponding parameter. We have studied a lot of examples of just how you go about doing that.

Various Destinations

The topics that are not included in this book are very similar to the ones that are. They merely involve different (and usually more complicated) populations, parameters, samples, statistics, and sampling distributions.

One direction in which you might consider going will lead you to the *general linear model* that includes regression analyses, *t* tests, and the analysis of variance. The inferential aspects of the general linear model are subsumed under the heading of *parametric statistics*, because certain assumptions about the population distributions and their parameters are made, such as the equality of population variances when testing the significance of the difference among several sample means.

Another direction leads to *non-parametric statistics*. These inferential statistical tests make fewer or no assumptions about the population distributions and their parameters and are often called *distribution-free*. You will find that such tests are very easy to carry out. The Kruskal-Wallis counterpart to "one-way" analysis of variance is a typical example of a non-parametric test of significance.

As this book draws to a close, I feel I should have some additional words of wisdom to pass along to you as you prepare to go forth to meet the cruel statistical world. All I can think of to say, however, is that if you really understand the five terms listed at the beginning of this chapter you will always know what statistics is all about. The converse unfortunately also holds: If you do not understand those five terms you will never know what it's all about. I hope and pray that you fall into the former category. Good luck!

An Annotated Bibliography of Recommended Statistics Books

Against all odds [video series]. (1990). Santa Barbara, CA: Intellimation.

> *This is a series of 26 videotaped programs on various topics in statistics (available from Intellimation, P.O. Box 1922, Santa Barbara, CA 93116-1922). The audiovisual aspects of the series are impressive, and the examples are varied and interesting, but there are several errors in the statistical content of which you should be aware (see the review of the series by Gabriel et al. in the November 1990 issue of the* American Statistician.*)*

Agresti, A., & Finlay, B. (1986). *Statistics for the social sciences* (2nd ed.). New York: Dellen.

> *This is an excellent text for students who are concentrating in sociology, psychology, or any of the social sciences.*

Note: Some of the books listed may strike you as a bit old; those that were published more than 10 years ago are considered classics in the field and/or are particular favorites of mine.

Cohen, J. (1988). *Statistical power analysis for the behavioral sciences* (2nd ed.). Hillsdale, NJ: Lawrence Erlbaum.

This book provides a lucid discussion of the concept of power, along with a variety of formulas and tables for determining the appropriate sample size for a number of common hypothesis testing procedures.

Darlington, R. (1990). *Regression and linear models*. New York: McGraw-Hill.

This is one of the best textbooks available concerning regression analysis and the general linear model. The inexpensive statistical software MYSTAT is referred to throughout this book to illustrate the various computations.

Fleiss, J. (1981). *Statistical methods for rates and proportions* (2nd ed.). New York: John Wiley.

This is the only statistics book I know of (other than this one) that concentrates on rates, proportions, and, of course, percentages. The mathematics gets heavy at times, but the results make working through the notation and the formulas well worth the effort.

Freedman, D., Pisani, R., Purves, R., & Adhikari, A. (1991). *Statistics* (2nd ed.). New York: W. W. Norton.

This very popular text provides the reader with a thorough grounding in basic concepts. The illustrations are particularly informative and often hilarious.

Hodges, J. (1975). *Statlab: An empirical introduction to statistics*. New York: McGraw-Hill.

Hodges's book requires total involvement on the part of the student in drawing samples, interpreting results, and so on. (You should be used to that by now.) The book comes complete with colored dice for probability and sampling exercises and a delightful set of real data gathered in a medical survey.

Huff, D. (1954). *How to lie with statistics*. New York: W. W. Norton.

This volume is a very amusing, but also very informative, spoof of statistics.

Huff, D. (1959). *How to take a chance*. New York: W. W. Norton.

This spoof of probability is comparable to Huff's spoof of statistics in How to Lie With Statistics.

Jaeger, R. (1990). *Statistics: A spectator sport* (2nd ed.). Newbury Park, CA: Sage.

This splendidly written text discusses the basic concepts of statistics, measurement, and research design, as well as a number of advanced statistical techniques (e.g., the analysis of variance and the analysis of covariance) near the end of the book. (And it also has no formulas!)

Kanji, G. K. (1993). *100 statistical tests*. Newbury Park, CA: Sage.

Once you feel comfortable that you understand basic statistical concepts, this book is a good place to turn for brief discussions and examples of many of the most common hypothesis testing procedures encountered in the research literature. There are a few unfortunate typographical errors, however, that mar this otherwise excellent reference text.

MacNeal, E. (1994). *Mathsemantics.* New York: Viking.

As the title implies, this volume is concerned with both the doing and the meaning of mathematics. The chapter on percentages and how poorly they are understood by the general populace is particularly interesting.

Paulos, J. A. (1995). *A mathamatician reads the newspaper.* New York: Basic Books.

Paulos has a few particularly good chapters that deal with statistics.

Siegel, S., & Castellan, J. (1988). *Non-parametric statistics for the behavioral sciences* (2nd ed.). New York: McGraw-Hill.

This revised "cookbook classic" contains descriptions and examples of all of the popular nonparametric tests of statistical significance.

Vogt, W. P. (1993). *Dictionary of statistics and methodology.* Newbury Park, CA: Sage.

This is a good source for definitions of statistical concepts that you may come across and that are not covered in this book.

Answers to
(Most of) the Exercises

Note: Some of the exercises do not have "right" answers, because the answer depends upon which cards are actually drawn.

Chapter 1

1. Ethel Eaglestone, Joseph Juzwiak, and Richard Rzepkowski.
   Yes, that would be a dichotomy, because it is a variable that has just two categories ("same" and "different").
   It would be *very* skewed, because the frequency for "same" is 3 and the frequency for "different" is 49.
2. Highest (a score of 13): Nancy Gabbey, Richard Mytych, Luigi Tabacco, and Alice Zutterman.
   Lowest (a score of 1): Donna Abbey, David Gzik, Bruce Naab, and Peggy Tylter.

3. Most letters (18): District of Columbia.

   Fewest letters (4): A three-way tie among Iowa, Utah, and Ohio. The frequency distribution for number of letters in name is as follows:

Value	Tally	Frequency	Relative Frequency
4	111	3	.058
5	111	3	.058
6	11111	5	.096
7	111111111	9	.173
8	11111111111	11	.212
9	111111	6	.115
10	111	3	.058
11	11111	5	.096
12	111	3	.058
13	111	3	.058
14		0	.000
15		0	.000
16		0	.000
17		0	.000
18	1	1	.019
		52	

As you can see, this distribution is positively skewed, because of the long "tail" to the right (if you rotate your book 90 degrees counter-clockwise), even though it is fairly symmetric for values between 4 and 13. The District of Columbia (18 letters) is an *outlier*.

4a. Variable = number of members in the U.S. House of Representatives

Value	Tally	Frequency	Relative Frequency
0- 4	111111111111111111111	21	.404
5- 9	11111111111111111	17	.327
10-14	111111	6	.115
15-19	11	2	.038
20-24	111	3	.058
25-29		0	.000
30-34	11	2	.038
35-39		0	.000
40-44		0	.000
45-49		0	.000
50-54	1	1	.000
		52	

(*Note:* It was necessary to group certain numbers together (0-4, 5-9, and so on) in order that the frequency distribution would not look too "anemic." This distribution is even more positively skewed than the distribution for Exercise 3. Blame it on California, which is the outlier this time.)

4b. 25; that is, just less than half. (I'll bet you thought it was many more than half, didn't you?)

5. If you know the answer to this one, you should run for president of the United States (or for God)!

Chapter 2

1. Mean = 7; standard deviation = 3.742. You didn't have to do any calculations, did you (because the vocabulary variable is "the same" as the denomination variable)?

2a. 12 (from 1 to 13).

2b. One-half of the range is $1/2 \times 12 = 6$. The range divided by the square root of twice the number of observations is 12 divided by the square root of 104; that is, 12 divided by 10.198, or 1.177. As 3.742 is between 1.177 and 6, we know that we're at least in the right ballpark.

3. Mean = 8.365; standard deviation = 9.414.

4. Skewness = 2.464; kurtosis = 10.427. The skewness makes sense, because of the long tail to the right "toward California." The kurtosis seems awfully high, but is undoubtedly due to both the peak at the left and the long tail to the right. (See the frequency distribution for Exercise 4, Chapter 1.) If you did all of these calculations by hand you may have noticed how much effect California had on the skewness and the kurtosis, and how little effect all the other states had.

5. It would be like comparing apples and oranges. For example, if we know that the mean height of a population of adult males is 69 inches, with a standard deviation of 3 inches, and their mean weight is 170 pounds, with a standard deviation of 15 pounds, that's all we know. We can't say that they are heavier than they are tall, even though the mean weight of 170 is a bigger number than the mean height of 69, because weight is measured in pounds and height is measured in inches. Similarly, we can't use the two standard deviations of 3 inches and 15 pounds to argue that they are more spread out in weight (pardon the pun) than they are in height, for the same reason. (There is actually another statistic, called the *coefficient of variation*, that does permit this sort of interpretation.)

Chapter 3

1. $3/52 = .058 = 5.8\%$.
2. Because the mean is 7 and the standard deviation is 3.742, all children with scores of 11 or higher are more than one standard deviation above the mean. There are 4 who scored 11, 4 who scored 12, and 4 who scored 13, for a total of 12 children out of 52, or 23.1%.
3. 9 out of 52, or 17.3%. (See the frequency distribution for Exercise 3, Chapter 1.)
4. Referring to the actual data in Table 1.6, 2 states have no representatives, 7 states have 1, and 12 have 11 or more, for a total of 21 out of 52, or 40.4%. There could be a rounding problem here if you calculated the percentages separately for no representatives, 1 representative, 11 representatives, and so on, and then added those percentages.
5. 25 out of 52, or 48.1%.

Chapter 4

1. 1/2, or .50, because there are 26 girls out of 52 children (26/52 reduces to 1/2, which is the same as .50).

   51/52, or .981. This can also be obtained by calculating the probability of selecting Donna Abbey (1/52, or .019) and subtracting that from 1.
2. $26/52 \times 25/51$, or .245.
3. The probability of getting all girls is $1/2 \times 1/2 \times 1/2 = 1/8$, or .125. The probability of that *not* happening is $1 - 1/8 = 7/8$, or .875. Therefore, the odds against getting all girls is 7/8 divided by 1/8, or .875/.125—that is, 7 to 1.
4. This is a tough question. If order of selection is important in defining "different samples," the answer is $52 \times 51 \times 50 \times 49 = 6,497,400$ (permutations). If order is not important in defining "different samples," then there are "only" $52 \times 51 \times 50 \times 49$ *divided by* $1 \times 2 \times 3 \times 4$; that is, 6,497,400/24, or 270,725 (combinations). In either event, I'll bet that's a lot more than you thought it would be.
5. 14/52, or .269.

Chapter 5

2. It would get "skinnier," because samples of five each are more likely to represent the population than are samples of two each, and therefore the statistics based on the larger sample size are likely to vary less from one another.

3. The general shape of the sampling distribution would be similar, but it would be "fleshed out" better because taking 100 samples rather than 50 samples would provide a better fit to the theoretical sampling distribution for that statistic.

5. The standard error is best interpreted as the typical amount by which an obtained sample statistic is expected to differ from the corresponding population parameter.

Chapter 6

2. The statistic based on the combined sample of 60 observations *should* be closest, because a sample of 60 takes a bigger chunk out of the population than a sample of 30, but "by chance" it may not.

4. A .68 confidence interval would be narrower, because you "lay off" only one standard error; a .95 confidence interval for a sample of 60 would also be narrower, because the standard error would be smaller, so there would be a smaller quantity to lay off on either side.

Chapter 7

3. The correctness of the inference regarding the difference between two percentages does not depend upon the correctness of the inferences regarding the hypothesized percentage for each of them separately. For example, sample percentages of 40 and 60 might be equally commensurate with a hypothesized percentage of 50, but the difference between 40 and 60 might not be commensurate with a hypothesized percentage difference of 0.

5. Whether a sample size of 30 was "appropriate" depends entirely upon your alternative hypothesis, your chosen significance level, and your desired power. If your alternative hypo-

thesis postulated a "big effect" (i.e., the parameter hypothesized in the alternative hypothesis was quite different from the parameter hypothesized in the null hypothesis), your chosen significance level was "liberal" (e.g., .05 as opposed to .01), and your desired power was not too high (.80, say, as opposed to .95), then a sample size of 30 is perfectly fine (and might even be *too* large). But if you had very stringent specifications (e.g., an alternative close to the null, the .01 level of significance, and desired power of .95), 30 is much too small.

Chapter 8

3. For a sample size of 200 (as opposed to 20), you would be equally likely to make a Type I error, because you *specify* that before you see the data, but you would be less likely to make a Type II error, because you would have greater power. The relative risk and the odds ratio would be unaffected, because all of the frequencies would be multiplied by 10, and things would cancel out.

Index of Terms

About the Author

Thomas R. Knapp, Ed.D., is Professor of Nursing and Education at The Ohio State University. His specialty is quantitative methodology (statistics, measurement, and research design). He is coauthor (with Bethel Ann Powers) of *A Dictionary of Nursing Theory and Research* (Sage, 1995) and has published several articles in nursing and educational research journals. His most recent articles include "Regression Analyses: What to Report" (*Nursing Research,* 1994); "The National Survey of Families and Households: A Rich Data Base for Nursing Research" (*Research in Nursing & Health,* 1995; coauthored with Teresa W. Julian); and "Ten Measurement Commandments That Often Should Be Broken" (*Research in Nursing & Health,* 1995; coauthored with Jean K. Brown). He has made several presentations at educational and nursing research conferences, and was the Gladys E. Sorenson Distinguished Lecturer in 1994 (his lecture was titled "Validity and Reliability: A Minority View").